TUTORIAL CHEMISTRY TEXTS

5
Structure and Bonding

JACK BARRETT

Imperial College of Science, Technology and Medicine, University of London

RS•C
ROYAL SOCIETY OF CHEMISTRY

Cover images © Murray Robertson/visual elements 1998–99, taken from the
109 Visual Elements Periodic Table, available at www.chemsoc.org/viselements

ISBN 0-85404-647-X

A catalogue record for this book is available from the British Library

Published by The Royal Society of Chemistry, Thomas Graham House, Science Park,
Milton Road, Cambridge CB4 0WF, UK
Registered Charity No. 207890
For further information see our web site at www.rsc.org

Typeset in Great Britain by Wyvern 21, Bristol
Printed and bound by Polestar Wheatons Ltd, Exeter

Preface

This book on the structure of molecules and their bonding is intended for the introductory courses given in the first two years of chemistry courses in UK universities. The subject is a complex one, and as Robert S. Mulliken, winner of the 1966 Nobel prize for his contribution to chemical bonding known as molecular orbital theory, said: "the chemical bond is not a simple as some people seem to think it is". In this book it is recognized that the solutions to quantum mechanical problems are the foundation of the subject. It is the author's opinion that undergraduates should be exposed to as much of atomic and molecular theory as is compatible with their incoming qualifications and the requirements for the other parts of their chemistry courses. The more understanding of the basic theory they acquire, the more will their appreciation be of chemistry in general.

The identification of a chemical bond with a pair of electrons came from G. N. Lewis in 1916, and the term covalent bond was coined by Irving Langmuir. The Lewis structures for covalent bonds are still in use in the UK A-level syllabus, and it is the main purpose of this book to persuade new students that the time has come to make further progress with their understanding of bonding.

This book has evolved from the author's previous attempts to give a general description of the ideas which are used to describe atomic and molecular structure, and fulsome thanks go to Ellis Horwood, the publisher of some of my books, for his generous permission to use some of the material in the present volume.

Martyn Berry took on the onerous task of reading every word of the manuscript, and thanks go to him for the many suggestions that produced an improved final text.

Jack Barrett
London

TUTORIAL CHEMISTRY TEXTS

EDITOR-IN-CHIEF

Professor E W Abel

EXECUTIVE EDITORS

Professor A G Davies
Professor D Phillips
Professor J D Woollins

EDUCATIONAL CONSULTANT

Mr M Berry

This series of books consists of short, single-topic or modular texts, concentrating on the fundamental areas of chemistry taught in undergraduate science courses. Each book provides a concise account of the basic principles underlying a given subject, embodying an independent-learning philosophy and including worked examples. The one topic, one book approach ensures that the series is adaptable to chemistry courses across a variety of institutions.

Further information about this series is available at www.chemsoc.org/tct

Orders and enquiries should be sent to:
Sales and Customer Care, Royal Society of Chemistry, Thomas Graham House, Science Park, Milton Road, Cambridge CB4 0WF, UK

Tel: +44 1223 432360; Fax: +44 1223 423429; Email: sales@rsc.org

Contents

1

A Brief Summary of Atomic Theory, the Basis of the Periodic Table and Some Trends in Atomic Properties

As a preliminary to the main topics of this book – molecular structure and chemical bonding – this chapter is included to remind the reader of the foundations of atomic theory, an understanding of which is necessary for the assimilation of the main subject matter.

Aims

From your previous studies and by the end of this chapter you should understand:

- Heisenberg's uncertainty principle and the necessity for quantum mechanics in the study of atomic structure
- What is meant by an atomic orbital
- The quantum rules for describing atomic orbitals
- The spatial orientations of s, p and d atomic orbitals
- Boundary surfaces and atomic orbital envelope diagrams
- That electrons possess an amount of intrinsic energy governed by the spin quantum number
- The Pauli exclusion principle
- That the numbers of atomic orbitals in an atom are dependent upon the Pauli exclusion principle
- That a maximum of two electrons may occupy an atomic orbital
- The *aufbau* principle and the order of filling the available atomic orbitals
- Hund's rules for the filling of degenerate orbitals
- The general structure of the Periodic Table
- The periodicity in first ionization energies of the elements
- The periodicity in electron attachment energies of the elements
- The definitions of atomic size

- The variations in atomic sizes across periods and down groups
- The definitions of electronegativity coefficients
- The variations of electronegativity coefficients across periods and down groups

1.1   The Heisenberg Uncertainty Principle

The uncertainty principle rationalizes our inability to observe the momentum and position of an atomic particle simultaneously. The act of observation interferes with atomic particles so that their momenta and positions are altered.

Heisenberg was awarded the 1932 Nobel Prize for Physics.

The Heisenberg Uncertainty Principle or principle of indeterminacy may be stated in the form: *It is impossible to determine simultaneously the position and momentum of an atomic particle*. It can also be stated by the simple mathematical relationship:

$$\Delta p \Delta q \approx h \tag{1.1}$$

in which Δp represents the uncertainty in momentum of the electron and Δq its uncertainty in position. The main consequence of the uncertainty principle is that because electronic energy levels are known with considerable accuracy, the positions of electrons within atoms are not known at all accurately. This realization forces theoretical chemistry to develop methods of calculation of electronic positions in terms of probabilities rather than assigning to them, for example, fixed radii around the nucleus. The varied methods of these calculations are known collectively as **quantum mechanics**.

1.2   Atomic Orbitals

Atomic orbitals represent the locations of electrons in atoms, and are derived from quantum mechanical calculations. The consequence of being in considerable ignorance about the position of an electron in an atom is that calculations of the **probability of finding an electron** in a given position must be made.

1.3   Quantum Mechanics

The solution of the **Schrödinger equation** gives mathematical form to the **wave functions** which describe the locations of electrons in atoms. The wave function is represented by ψ, which is such that its square, ψ^2, is the **probability density** of finding an electron.

The value of $\psi^2 d\tau$ represents the probability of finding the electron in the volume element, $d\tau$ (which may be visualized as the product of three elements of the Cartesian axes: $dx.dy.dz$). Each solution of the Schrödinger wave equation is known as an **atomic orbital**. Although the solution of the Schrödinger equation for any system containing more than one electron requires the iterative techniques available to computers, it may be solved for the hydrogen atom (and for hydrogen-like atoms such as He^+, Li^{2+}, *etc.*) by analytical means, the energy solutions being represented by the equation:

$$E_n = -\frac{N_A \mu Z^2 e^4}{8\varepsilon_0^2 h^2}\left[\frac{1}{n^2}\right]$$

(1.2)

where Z is the atomic number, μ is the reduced mass of the system, e is the electronic charge, and ε_0 is the **permittivity** of a vacuum; n is the **principal quantum number**. Equation 1.2 indicates that the energies of atomic orbitals increase as the value of n^2 increases, *i.e.* the electrons become less stable as the value of the principal quantum number increases.

The **quantum rules** describe atomic orbitals. The permitted values of the **principal quantum number**, n, are: $n = 1, 2, 3, 4, \ldots$; those of the **orbital angular momentum quantum number**, l, are: $l = 0, 1, 2, 3, \ldots$ $(n-1)$; and those of the **magnetic quantum number**, m_l, are: $m_l = +l, +l - 1, \ldots 0, \ldots -(l-1), -l$. Code letters for atomic orbitals, s, p, d and f, are used to represent those orbitals for which the value of l is 0, 1, 2 and 3, respectively.

Worked Problems

Q Is the combination of quantum number values $n = 4$, $l = 3$ and $m_l = 4$ a valid one?

A No, because the value of m_l must not exceed that of l.

Q Is the combination of quantum number values $n = 5$, $l = 4$ and $m_l = 4$ a valid one?

A Yes, but the combination refers to 5g orbitals which are not used by elements in their ground states.

The mathematical functions representing atomic orbitals are usually **normalized**, *i.e.* the wave function is arranged so that the integral of its square over all space has the value $+1$, as expressed by the general equation:

$$\int_0^\infty \psi^2 d\tau = 1 \qquad (1.3)$$

The solutions of the Schrödinger equation show how ψ is distributed in the space around the nucleus of the hydrogen atom. The solutions for ψ are characterized by the values of three quantum numbers and every allowed set of values for the quantum numbers, together with the associated wave function, strictly defines that space which is termed an atomic orbital. Other representations are used for atomic orbitals, such as the **boundary surface** and **orbital envelopes** described later in the chapter.

1.4   Spatial Orientations of Atomic Orbitals

The **spatial orientations** of the atomic orbitals of the hydrogen atom are very important in the consideration of the interaction of orbitals of different atoms in the production of **chemical bonds**.

Although the formal method of describing orbitals is to use mathematical expressions, much understanding of orbital properties may be gained by the use of pictorial representations. The most useful pictorial representations of atomic orbitals are similar to boundary surfaces (which are based on ψ^2), but are based upon the distribution of ψ values, with the sign of ψ being indicated in the various parts of the diagram. The shapes of these distributions are based upon the contours of ψ within

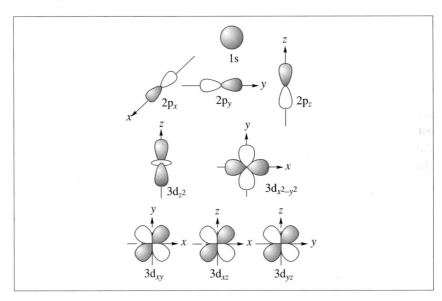

Figure 1.1 The envelope diagrams of the 1s, 2p ad 3d atomic orbitals of the H atom

which the values of ψ^2 represent 0.95, and may be called **orbital envelope diagrams**. The orbital envelopes for the 1s, three 2p and five 3d hydrogen orbitals are shown in Figure 1.1.

Two methods are used to represent the positive and negative values of the wave functions in **atomic orbital envelope diagrams**. Both are shown in Figure 1.2 for the $2p_z$ orbital. The two lobes of the orbital may be indicated by open areas containing the signs of the wave function in the two areas, or they may be represented by filled and open areas which represent the positive and negative values of the wave function, respectively. In the remainder of the book, the filled and open parts of atomic orbital diagrams are used to denote positive and negative signs of wave functions.

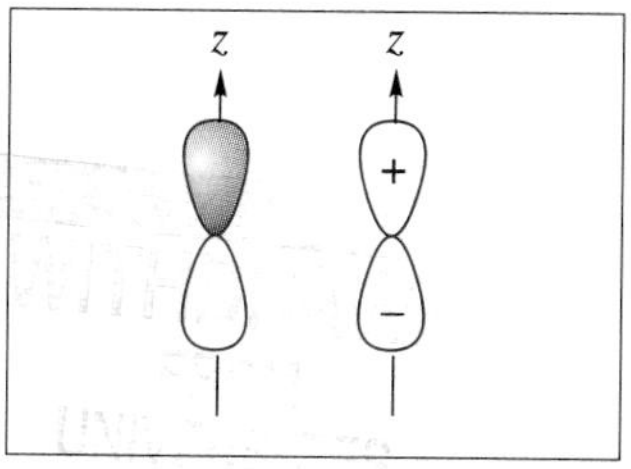

Figure 1.2 Two methods of representing the 2p orbitals

1.5 The Electronic Configurations of Atoms; the Periodic Classification of the Elements

The chemistry of an element is determined by the manner in which its electrons are arranged in the atom. Such arrangements are the basis of the modern periodic classification of the elements: the **Periodic Table**.

The treatment of atoms with more than one electron (**polyelectronic atoms**) requires consideration of the effects of **interelectronic repulsion**, **orbital penetration** towards the nucleus, **nuclear shielding**, and an extra quantum number (the **spin quantum number**) which specifies the **intrinsic energy** of the electron in any orbital. The restriction on numbers of atomic orbitals and the number of electrons that they can contain leads to a discussion of the **Pauli exclusion principle**, **Hund's rules** and the *aufbau* principle. All these considerations are necessary to allow the construction of the modern form of the periodic classification of the elements.

1.5.1 The Spin Quantum Number

When dealing with atoms possessing more than one electron it is necessary to introduce a fourth quantum number: s, the **spin quantum number**. However, to take into account the intrinsic energy of an electron, the value of s is taken to be $\frac{1}{2}$. Essentially the intrinsic energy of the electron may interact in a quantized manner with that associated with the angular momentum represented by l, such that the only permitted interactions are $l + s$ and $l - s$. For atoms possessing more than one electron it is necessary to specify the values of s with respect to an applied magnetic field; these are expressed as values of m_s of $+\frac{1}{2}$ or $-\frac{1}{2}$.

1.5.2 The Pauli Exclusion Principle

The Pauli exclusion principle is the cornerstone in understanding the chemistry of the elements. It may be stated as: "*No two electrons in an atom may possess identical sets of values of the four quantum numbers, n, l, m_l and m_s*". The consequences are:

1. To restrict the number of electrons per orbital to a maximum of two.
2. To restrict any one atom to only one particular orbital, defined by its set of n, l and m_l values.

Worked Problems

Q How many electrons in a given atom may possess the set of values of the three quantum numbers $n = 3$, $l = 2$ and $m_l = 1$?

A Two, because the only ways of producing unique values of the four quantum numbers (*i.e.* including m_s) are those of the three in the question associated with the two permitted m_s values of $+\frac{1}{2}$ and $-\frac{1}{2}$.

Q How many 2p electrons may there be in atoms (except those in the first short period)?

A Six, because there are three possible values of the m_l quantum number, all being associated with the two permitted m_s values of $+\frac{1}{2}$ and $-\frac{1}{2}$.

1.5.3 The Electronic Configurations and Periodic Classification of the Elements

The modern form of the Periodic Table is shown in Figure 1.3. The atomic number, Z, of each element is shown together with its electronic configuration. There are 18 **groups**, according to modern convention. The quantum rules define the different types of atomic orbitals which may be used for electron occupation in atoms. The Pauli exclusion principle defines the number of each type of orbital and limits each orbital to a maximum electron occupancy of two. Experimental observation, together with some sophisticated calculations, indicates the energies of the available orbitals for any particular atom. The **electronic configuration** (the orbitals which are used to accommodate the appropriate number of electrons) may be decided by the application of what is known as the *aufbau* (German: building-up) **principle**, which is that electrons in

1	2	3	4	5	6	7	8	9	10	11	12	13	14	15	16	17	18
1 **H** $1s^1$																	2 **He** $1s^2$
3 **Li** $2s^1$	4 **Be** $2s^2$											5 **B** $2s^2 2p^1$	6 **C** $2s^2 2p^2$	7 **N** $2s^2 2p^3$	8 **O** $2s^2 2p^4$	9 **F** $2s^2 2p^5$	10 **Ne** $2s^2 2p^6$
11 **Na** $3s^1$	12 **Mg** $3s^2$											13 **Al** $3s^2 3p^1$	14 **Si** $3s^2 3p^2$	15 **P** $3s^2 3p^3$	16 **S** $3s^2 3p^4$	17 **Cl** $3s^2 3p^5$	18 **Ar** $3s^2 3p^6$
19 **K** $4s^1$	20 **Ca** $4s^2$	21 **Sc** $4s^2 3d^1$	22 **Ti** $4s^2 3d^2$	23 **V** $4s^2 3d^3$	24 **Cr** $4s^1 3d^5$	25 **Mn** $4s^2 3d^5$	26 **Fe** $4s^2 3d^6$	27 **Co** $4s^2 3d^7$	28 **Ni** $4s^2 3d^8$	29 **Cu** $4s^1 3d^{10}$	30 **Zn** $4s^2 3d^{10}$	31 **Ga** $4s^2 3d^{10} 4p^1$	32 **Ge** $4s^2 3d^{10} 4p^2$	33 **As** $4s^2 3d^{10} 4p^3$	34 **Se** $4s^2 3d^{10} 4p^4$	35 **Br** $4s^2 3d^{10} 4p^5$	36 **Kr** $4s^2 3d^{10} 4p^6$
37 **Rb** $5s^1$	38 **Sr** $5s^2$	39 **Y** $5s^2 4d^1$	40 **Zr** $5s^2 4d^2$	41 **Nb** $5s^1 4d^4$	42 **Mo** $5s^1 4d^5$	43 **Tc** $5s^1 4d^6$	44 **Ru** $5s^1 4d^7$	45 **Rh** $5s^1 4d^8$	46 **Pd** $4d^{10}$	47 **Ag** $5s^1 4d^{10}$	48 **Cd** $5s^2 4d^{10}$	49 **In** $5s^2 4d^{10} 5p^1$	50 **Sn** $5s^2 4d^{10} 5p^2$	51 **Sb** $5s^2 4d^{10} 5p^3$	52 **Te** $5s^2 4d^{10} 5p^4$	53 **I** $5s^2 4d^{10} 5p^5$	54 **Xe** $5s^2 4d^{10} 5p^6$
55 **Cs** $6s^1$	56 **Ba** $6s^2$	57 **La** $6s^2 5d^1$	72 **Hf** $4f^{14} 6s^2 5d^2$	73 **Ta** $4f^{14} 6s^2 5d^3$	74 **W** $4f^{14} 6s^2 5d^4$	75 **Re** $4f^{14} 6s^2 5d^5$	76 **Os** $4f^{14} 6s^2 5d^6$	77 **Ir** $4f^{14} 6s^2 5d^7$	78 **Pt** $4f^{14} 6s^1 5d^9$	79 **Au** $4f^{14} 6s^1 5d^{10}$	80 **Hg** $4f^{14} 6s^2 5d^{10}$	81 **Tl** $4f^{14} 6s^2 5d^{10} 6p^1$	82 **Pb** $4f^{14} 6s^2 5d^{10} 6p^2$	83 **Bi** $4f^{14} 6s^2 5d^{10} 6p^3$	84 **Po** $4f^{14} 6s^2 5d^{10} 6p^4$	85 **At** $4f^{14} 6s^2 5d^{10} 6p^5$	86 **Rn** $4f^{14} 6s^2 5d^{10} 6p^6$
87 **Fr** $7s^1$	88 **Ra** $7s^2$	103 **Lr** $5f^{14} 7s^2 6d^1$	104 **Rf** $5f^{14} 7s^2 6d^2$	105 **Db** $5f^{14} 7s^2 6d^3$	106 **Sg** $5f^{14} 7s^2 6d^4$	107 **Bh** $5f^{14} 7s^2 6d^5$	108 **Hs** $5f^{14} 7s^2 6d^6$	109 **Mt** $5f^{14} 7s^2 6d^7$	110	111	112		114		116		118

Lanthanide elements

57 **La** $6s^2 5d^1$	58 **Ce** $4f^1 6s^2$	59 **Pr** $4f^3 6s^2$	60 **Nd** $4f^4 6s^2$	61 **Pm** $4f^5 6s^2$	62 **Sm** $4f^6 6s^2$	63 **Eu** $4f^7 6s^2$	64 **Gd** $4f^7 5d^1 6s^2$	65 **Tb** $4f^9 6s^2$	66 **Dy** $4f^{10} 6s^2$	67 **Ho** $4f^{11} 6s^2$	68 **Er** $4f^{12} 6s^2$	69 **Tm** $4f^{13} 6s^2$	70 **Yb** $4f^{14} 6s^2$

Actinide elements

89 **Ac** $7s^2 6d^1$	90 **Th** $7s^2 6d^2$	91 **Pa** $5f^2 7s^2 6d^1$	92 **U** $5f^3 7s^2 6d^1$	93 **Np** $5f^4 7s^2 6d^1$	94 **Pu** $5f^6 7s^2$	95 **Am** $5f^7 7s^2$	96 **Cm** $5f^7 7s^2 6d^1$	97 **Bk** $5f^9 7s^2$	98 **Cf** $5f^{10} 7s^2$	99 **Es** $5f^{11} 7s^2$	100 **Fm** $5f^{12} 7s^2$	101 **Md** $5f^{13} 7s^2$	102 **No** $5f^{14} 7s^2$

Figure 1.3 The Periodic Table showing the atomic numbers and outer electronic configurations of the elements; the core configurations are those of the preceding Group 18 element; those that are numbered, but unnamed, have been synthesized in small quantities

the ground state of an atom occupy the orbitals of lowest energy such that the total electronic energy is minimized.

The various considerations that allow the *aufbau* order of filling of atomic orbitals as the nuclear charge increases steadily along the order of the elements can be written down in a manner similar to the Periodic Table as shown in Table 1.1, starting on the top left-hand side with the 1s orbital and then filling each successive row, ending with the incompletely filled 7p orbitals. There are some irregularities in the filling, but the order of filling is generally in line with the value of the sum $n + l$, and if there is more than one set of orbitals with the same sum, the set with the larger or largest l value is filled preferentially. For example, the 3d and 4p sets of orbitals have a sum of $n + l = 5$, but the 3d orbitals are filled before the 4p.

Table 1.1 An arrangement showing the relationship between the atomic orbitals filled and the number of elements in the various groups and periods of the Periodic Table. Both the major blocks mimic the arrangement of the elements in the 18-group Periodic Table

Orbitals filled				Numbers of elements associated with the maximum filling of the orbitals			
1s				2			
2s			2p	2			6
3s			3p	2			6
4s		3d	4p	2		10	6
5s		4d	5p	2		10	6
6s	4f	5d	6p	2	14	10	6
7s	5f	6d	7p[a]	2	14	10	6

[a]Incomplete at this time

Notice the similarities between the two parts of Table 1.1 and the form of the Periodic Table given in Figure 1.3. The quantum rules, the Pauli exclusion principle and the *aufbau* principle combine to explain the general structure of the Periodic Table.

Hund's rules are used to determine the ground state of atoms which have electrons in degenerate orbitals, *i.e.* those with identical energy, such as the three 2p orbitals. These may be stated in the following way: in filling a set of degenerate orbitals the number of unpaired electrons is maximized, and such unpaired electrons will possess parallel spins.

Worked Problem

Q How many unpaired electrons are there in the ground state of the carbon atom?

A Two, because the electronic configuration of $1s^2 2s^2 2p^2$ indicates that there are two 2p electrons in the triply degenerate 2p orbitals and they will occupy two of these singly.

1.6 Periodicities of Some Atomic Properties

This section summarizes the variation, across the periods and down the groups of the Periodic Table, of (i) the ionization energies, (ii) the electron attachment energies (electron affinities), (iii) the atomic sizes and (iv) the electronegativity coefficients of the elements.

Box 1.1 The Relationship between Internal Energies and Enthalpies

The ionization energies and electron attachment energies used in this and other chapters are values of ΔU, *i.e.* changes of internal energy. To convert ionization energies to ionization enthalpy changes at 298 K requires the *addition* of $^5/_2 RT$, which amounts to 6.2 kJ mol^{-1}. Similarly, to convert electron attachment energies to enthalpy changes requires the *subtraction* of $^5/_2 RT$.

1.6.1 Periodicity of First Ionization Energies

The first ionization energy of an atom is the minimum energy required to convert one mole of the gaseous atom (in its ground – lowest energy – electronic state) into one mole of its gaseous unipositive ion:

$$A(g) \rightarrow A^+(g) + e^- \qquad (1.4)$$

The first ionization energies of the elements are plotted in Figure 1.4. There is a characteristic pattern of the values for the elements Li to Ne which is repeated for the elements Na to Ar, and which is repeated yet again for the elements K, Ca and Al to Kr (the s- and p-block elements of the fourth period). In the latter case, the pattern is interrupted by the values for the 10 transition elements of the d-block. The fourth period pattern is repeated by the fifth period elements, and there is an additional section in the sixth period where the 14 lanthanide elements occur. In addition to this periodicity there is a general downward trend, and the ionization energy decreases down any of the groups. Both observations are broadly explicable in terms of the electronic configurations of the elements.

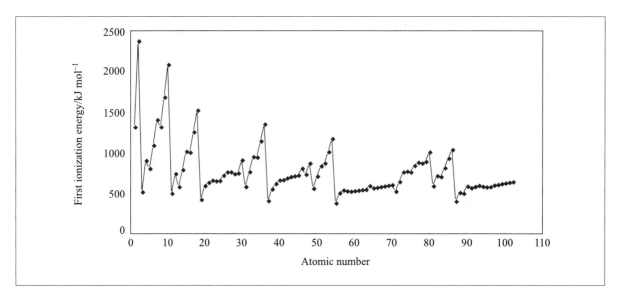

Figure 1.4 The first ionization energies of the elements

1.6.2 Variations in Electron Attachment Energies (Electron Affinities)

The **electron attachment energy** or **electron affinity** is defined as the *change* in internal energy *(i.e. ΔU)* that occurs when one mole of gaseous atoms of an element are converted by electron attachment to give one mole of gaseous negative ions:

$$A(g) + e^- \rightarrow A^-(g) \tag{1.5}$$

Take care to note the definition of electron affinity given in the text. In some older texts the term is defined as 'the energy *released* upon attachment of an electron to a neutral atom.' Using this definition the values given in this text would have the opposite sign.

Conversion to an electron attachment enthalpy requires the subtraction of $^5/_2RT$, *i.e.* 6.2 kJ mol^{-1} at 298 K. Most elements are sufficiently electronegative to make their *first* electron attachment energies *negative* according to the thermodynamic convention that *exothermic reactions are associated with negative energy changes*. All second and subsequent electron attachment energies are highly endothermic.

The first electron attachment energies of the first 36 elements are plotted in Figure 1.5 and show the values for H and He followed by a characteristic pattern, the second repetition of which is split by the values for the 10 transition elements. The value for hydrogen is –72.8 kJ mol^{-1}, which is very different from the 1s orbital energy of –1312 kJ mol^{-1} because of the interelectronic repulsion term amounting to –72.8 –

$(-1312) = 1239.2$ kJ mol^{-1}. The value for the helium atom is positive (21 kJ mol^{-1}), indicating an overall repulsion of the incoming electron from the $1s^2$ configuration which is not offset by the nuclear charge.

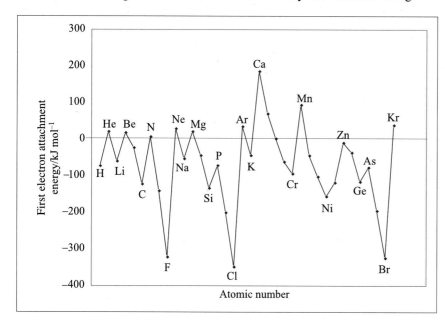

Figure 1.5 The electron attachment energies of the first 36 elements

1.6.3 Variations in Atomic Size

The size of an atom is not a simple concept. An inspection of the wave function for any atom shows that it is asymptotic to infinity, so some practical definition of size is required. There are two ways of assigning sizes to atoms: atomic radius and covalent radius.

All the atomic orbital wave functions contain the exponential term $e^{-\rho}$, where $\rho = Z_{eff}r/a_0$, which is zero only when $r = \infty$.

Atomic Radius

The atomic radius of an element is considered to be half the interatomic distance between identical (singly bonded) atoms. This may apply to iron, say, in its metallic state, in which case the quantity may be regarded as the metallic radius of the iron atom, or to a molecule such as Cl_2. The difference between the two examples is sufficient to demonstrate that some degree of caution is necessary when comparing the atomic radii of different elements. It is best to limit such comparisons to elements with similar types of bonding, metals for example. Even that restriction is subject to the drawback that the metallic elements have at least three different crystalline arrangements with possibly different **coordination numbers** (the number of nearest neighbours for any one atom).

Covalent Radius

The covalent radius of an element is considered to be one half of the covalent bond distance of a molecule such as Cl_2 (equal to its atomic radius in this case), where the atoms concerned are participating in single bonding. Covalent radii for participation in multiple bonding are also quoted in data books. In the case of a single bond between two different atoms, the bond distance is divided up between the participants by subtracting the covalent radius of one of the atoms, whose radius is known, from it. A set of mutually consistent values is now generally accepted and, since the vast majority of the elements take part in some form of covalent bonding, the covalent radius is the best quantity to consider for the study of general trends. Only atoms of the Group 18 elements (except Kr and Xe) do not have covalent radii assigned to them because of their general inertness with respect to the formation of molecules. The use of covalent radius for comparing the sizes of atoms is subject to the reservation that its magnitude, for any given atom, is dependent upon the oxidation state of that element.

Figure 1.6 shows how covalent radii vary across periods and down the main groups of the Periodic Table. Across periods there is a general reduction in atomic size, while down any group the atoms become larger. These trends are consistent with the understanding gained from the study of the variations of the first ionization energies of the elements. As the ionization energy is a measure of the effectiveness of the nuclear charge in attracting electrons, it might be expected that an increase in nuclear effectiveness would lead to a reduction in atomic size. The trends in atomic size in the periodic system are, in general, almost the exact opposite to those in the first ionization energy.

Note the relatively large reduction in covalent radius between the first and second members of each main group. This is consistent with the greater binding experienced by the 2s and by the 2p electrons in particular of the elements of the second period.

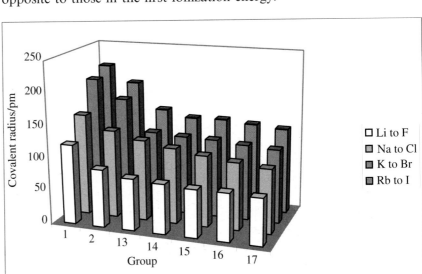

Figure 1.6 The covalent radii of the main group elements of the 2nd, 3rd, 4th and 5th periods

The details of size variations within the transition series and the lanthanide and actinide series are not included here, but are discussed in more specialist books.

1.6.4 Electronegativity

The concept of electronegativity is derived from experimental observations, such as that the elements fluorine and chlorine are highly electronegative in their strong tendencies to become negative ions. The metallic elements of Group 1, on the other hand, are not electronegative and are better described as being electropositive: they have a strong tendency to form positive ions. A scale of electronegativity coefficients is useful in allowing a number to represent the tendency of an element in a molecule of attracting electrons to itself. The establishment of such a scale has involved the powers of two Nobel prize winners: Pauling (chemistry prize for valence bond theory, 1954) and Mulliken (chemistry prize for molecular orbital theory, 1966). Both the Pauling and Mulliken scales suffered from lack of accurate data and have been largely replaced by the Allred–Rochow scale.

The Allred–Rochow Scale of Electronegativity Coefficients

The now generally accepted scale of electronegativity was derived by Allred and Rochow and is known by their names. It is based on the concept that the electronegativity of an element is related to the force of attraction experienced by an electron at a distance from the nucleus equal to the covalent radius of the particular atom.

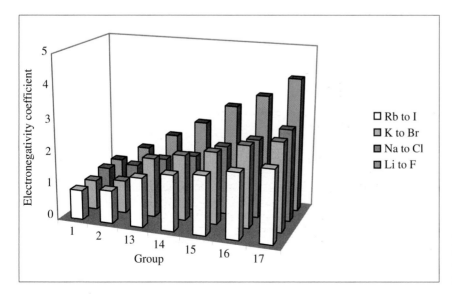

Figure 1.7 The Allred–Rochow electronegativity coefficients of the main group elements of the 2nd, 3rd, 4th and 5th periods

Figure 1.7 shows the variation of Allred–Rochow electronegativity coefficients for singly bonded elements along periods and down groups of the Periodic Table. In general the value of the electronegativity coefficient increases across the periods and decreases down the groups. That is precisely the opposite of the trends in covalent radii, but similar to the trends in first ionization energy (Figure 1.4). This latter conclusion is no surprise, as it is to be expected that elements with a high tendency to attract electrons possess high first ionization energies.

Summary of Key Points

This chapter is a reminder of the important aspects of

1. The theory of atomic orbitals including the definition of atomic orbitals by the three quantum numbers, n, l and m_l.
2. The general structure of the Periodic Table, based on atomic orbital energies, the *aufbau* principle, the Pauli exclusion principle and Hund's rules.
3. The trends in first ionization energies, first electron attachment energies, atomic sizes and electronegativity coefficients of the elements across the groups and down the periods of the periodic classification.

Problems

1. State whether the following sets of quantum number values are valid descriptions of atomic orbitals, and explain why some are invalid.

	n	l	m_l
(a)	2	2	0
(b)	3	0	−1
(c)	3	1	−1
(d)	5	4	3
(e)	4	3	4
(f)	4	2	2

2. In a hydrogen atom what are the degeneracies of the 3p, 4d and 5d atomic orbitals?

3. Why do the values of successive ionization energies of an atom always show an increasing trend?

4. What *general* trends are noticeable across the Periodic Table in the values of: (a) the first ionization energies, (b) the first electron attachment energies, and (c) the covalent radii of the elements?

5. What *general* trends are noticeable down the groups of the Periodic Table in the values of (a) the first ionization energies, (b) the first electron attachment energies and (c) the covalent radii of the elements?

Further Reading

J. Barrett, *Atomic Structure and Periodicity*, Royal Society of Chemistry, Cambridge, 2001. This book is meant to act as preliminary reading for the present text, but covers the subject matter in a largely non-mathematical way. The theoretical basis of the Periodic Table is dealt with in considerable detail and is followed by discussions of the periodicities of the main physical and chemical properties of the elements.

B. Hoffmann, *The Strange Story of the Quantum*, 2nd edn., Dover, New York, 1959. A highly readable book which gives insights to the concepts of quantum theory.

G. Herzberg, *Atomic Spectra and Atomic Structure*, 2nd edn., Dover, New York, 1944. This is a classic text by the recipient of the 1971 Nobel Prize for Chemistry, but is still available and contains descriptions of atomic spectra and atomic structure which are "from the horse's mouth!"

T. P. Softley, *Atomic Spectra*, Oxford University Press, Oxford, 1994. A very concise account of the subject.

D. O. Hayward, *Quantum Mechanics for Chemists*, Royal Society of Chemistry, Cambridge, 2001. A companion book in the Tutorial Chemistry Texts series.

R. J. Puddephatt and P. K. Monaghan, *The Periodic Table of the Elements*, 2nd edn., Oxford University Press, Oxford, 1986. A concise description of the structure of the Periodic Table and a discussion of periodic trends of many physical and chemical properties of the elements.

D. M. P. Mingos, *Essential Trends in Inorganic Chemistry*, Oxford University Press, Oxford, 1998. A new look at trends in physical and chemical properties of the elements and their compounds. This is a general text that would cover the material of most university courses in inorganic chemistry apart from some specialized topics, but is recommended here for the sections on atomic and molecular structure.

2

Molecular Symmetry and Group Theory

The contents of this chapter are fundamental in the applications of molecular orbital theory to bond lengths, bond angles and molecular shapes, which are discussed in Chapters 3–6. This chapter introduces the principles of group theory and its application to problems of molecular symmetry. The application of molecular orbital theory to a molecule is simplified enormously by the knowledge of the symmetry of the molecule and the group theoretical rules that apply.

Aims

By the end of this chapter you should understand:

- What an element of symmetry is
- What a symmetry operation is
- The differences between horizontal, vertical and dihedral planes of symmetry
- What a point group is
- What a representation is
- The difference between reducible and irreducible representations
- How to multiply representations
- How to assign molecules to point groups

2.1 Molecular Symmetry

A mathematical **group** consists of a set of **elements** which are related to each other according to certain rules, outlined later in the chapter. The particular kind of elements which are relevant to the symmetries of molecules are **symmetry elements**. With each symmetry element there is an associated **symmetry operation**. The necessary rules are referred to where appropriate.

2.1.1 Elements of Symmetry and Symmetry Operations

There are seven **elements of symmetry** which are commonly possessed by molecular systems. These elements of symmetry, their notations and their related symmetry operations are given in Table 2.1.

Table 2.1 Elements of symmetry and their associated operations

Symmetry element	Symbol	Symmetry operation
Identity	E	Leave the molecule alone
Proper axis	C_n	Rotate the molecule by $360/n$ degrees around the axis
Horizontal plane	σ_h	Reflect the molecule through the plane which is perpendicular to the major axis
Vertical plane	σ_v	Reflect the molecule through a plane which contains the major axis
Dihedral plane	σ_d	Reflect the molecule through a plane which bisects two C_2 axes
Improper axis	S_n	Rotate the molecule by $360/n$ degrees around the improper axis and then reflect the molecule through the plane perpendicular to the improper axis
Inversion centre or centre of symmetry	i	Invert the molecule through the inversion centre

An element of symmetry is possessed by a molecule if, after the associated symmetry operation is carried out, the atoms of that molecule are not perceived to have moved. The molecule is then in an **equivalent configuration**. The individual atoms may have moved but only to positions previously occupied by identical atoms.

You could be looking at a particular molecule (or, more practically, a model of a molecule), and if you blinked when a symmetry operation was carried out so that you did not observe the operation, you would not know that the operation had occurred.

The Identity, E

The symmetry element known as the **identity**, and symbolized by E (or in some texts by I), is possessed by all molecules independently of their shape. The related symmetry operation of leaving the molecule alone seems too trivial a matter to have any importance. The importance of E is that it is essential, for group theoretical purposes, for a group to contain it. For example, it expresses the result of performing some operations twice, *e.g.* the double reflexion of a molecule in any particular plane of symmetry. Such action restores every atom of the molecule to its original position so that it is equal to the performance of the operation of leaving the molecule alone, expressed by E.

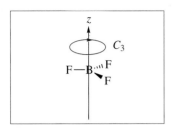

Figure 2.1 An example of a proper axis of symmetry, in this case of order 3, C_3

Proper Axes of Symmetry, C_n

A **proper axis of symmetry**, denoted by C_n, is an axis around which a molecule is rotated by $360°/n$ to produce an equivalent configuration. The trigonally planar molecule BF_3 may be set up so that the molecular plane is contained by the xy Cartesian plane (that containing the x and y axes) and so that the z Cartesian axis passes through the centre of the boron nucleus, as is shown in Figure 2.1.

If the molecule is rotated around the z axis by 120° (360°/3), an equivalent configuration of the molecule is produced. The boron atom does not change its position, and the fluorine atoms exchange places depending upon the direction of the rotation. The rotation described is the symmetry operation associated with the C_3 axis of symmetry, and the demonstration of its production of an equivalent configuration of the BF_3 molecule is what is required to indicate that the C_3 proper axis of symmetry is possessed by that molecule.

There are other proper axes of symmetry possessed by the BF_3 molecule. The three lines joining the boron and fluorine nuclei are all contained by C_2 axes (from hereon the term "proper" is dropped, unless it is absolutely necessary to remove possible confusion) as may be seen from Figure 2.2. The associated symmetry operation of rotating the molecule around one of the C_2 axes by 360°/2 = 180° produces an equivalent configuration of the molecule. The boron atom and one of the fluorine atoms do not move whilst the other two fluorine atoms exchange places. There are, then, three C_2 axes of symmetry possessed by the BF_3 molecule.

The value of the subscript n in the symbol C_n for a proper axis of symmetry is known as the **order** of that axis. The axis (and there may be more than one) of highest order possessed by a molecule is termed the **major axis**. The concept of the major axis is important in distinguishing between horizontal and vertical axes of symmetry. It is also important in the diagnosis of whether a molecule belongs to a C group or a D group (terms which are defined later in the chapter). As is the case with the C_n axis of BF_3, the axis of symmetry coincides with one of the Cartesian axes (z), but that is because of the manner in which the diagram in Figure 2.1 is drawn. It is a convention that *the major axis of symmetry should be coincident with the z axis*. It is unnecessary for there to be any coincidences between the axes of symmetry of a molecule and the Cartesian axes, but it is of considerable convenience if there is at least one coincidence.

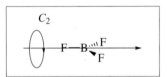

Figure 2.2 An example of a C_2 axis of symmetry

Planes of Symmetry, σ

There are three types of **planes of symmetry**, all denoted by the Greek lower case sigma, σ, and all of them are such that reflexions of the molecule through them produce equivalent configurations of that molecule. The symmetry operation associated with a plane of symmetry is reflexion through the plane.

(a) Horizontal Planes, σ_h. A **horizontal plane of symmetry** (denoted by σ_h) is one that is perpendicular to the major axis of symmetry of a molecule. The molecular plane of the BF_3 molecule is an example of a horizontal plane: it is perpendicular to the C_3 axis. Reflexion in the horizontal plane of the BF_3 molecule has no effect upon any of the four atoms. A better example is the PCl_5 molecule (shown in Figure 2.3) which, since it is trigonally bipyramidal, possesses a C_3 axis (again arranged to coincide with the z axis) and a horizontal plane (the xy plane as it is set up) which contains the phosphorus atom and the three chlorine atoms of the trigonal plane. Reflexion through the horizontal plane causes the apical (*i.e.* out-of-plane) chlorine atoms to exchange places in producing an equivalent configuration of the molecule.

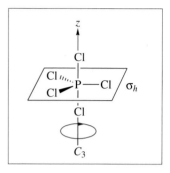

Figure 2.3 An example of a horizontal plane of symmetry, σ_h

Worked Problem

Q Are there other symmetry elements possessed by the PCl_5 molecule, other than the identity element, E?

A Yes, there are three C_2 proper axes of rotation that contain the lines joining the phosphorus atom with the three chlorine atoms in the horizontal plane and which are at 90° to the C_3 axis.

(b) Vertical Planes, σ_v. A **vertical plane of symmetry** (denoted by σ_v) is one which *contains* the major axis. The BF_3 and PCl_5 molecules both possess three vertical planes of symmetry. These contain the C_3 axis and, in the BF_3 case, the boron atom and one each of the fluorine atoms, respectively. In the PCl_5 case the three vertical planes, σ_v, contain the C_3 axis, the phosphorus atom, the two apical chlorine atoms and one each of the chlorine atoms of the trigonal plane, repectively.

Worked Problem

Q In the PCl_5 molecule, each of the three vertical planes contain another symmetry element. What is the element?

A Each vertical plane contains a C_2 proper axis of rotation.

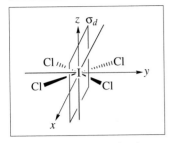

Figure 2.4 An example of a dihedral plane of symmetry, σ_d

(c) Dihedral Planes, σ_d. A **dihedral plane of symmetry** (denoted by σ_d) is one which *bisects two C_2 axes* of symmetry. In addition, it contains the major axis and so is a special type of vertical plane. An example is shown in Figure 2.4, which contains a diagram of the square planar ICl_4^- ion.

As may be seen from Figure 2.4, the ICl_4^- ion has a major axis which is a C_4 axis. The C_4 axis is also a C_2 axis in that rotations around it of 90° and 180° both produce equivalent configurations of the molecule. A rotation through 270° also produces an equivalent configuration, but that is equivalent to a rotation through 90° in the opposite direction and so does not indicate an extra type of axis of symmetry. The molecular plane contains two C_2' axes (the superscript prime is to distinguish them from the C_2 axis which is coincident with the C_4 axis), both of which contain the iodine atom and two diametrically opposed chlorine atoms. The two C_2' axes are contained respectively by the two vertical planes which also contain the major axis. There are, in addition, two C_2'' axes (coincident with the x and y axes) which are contained by the horizontal plane, are perpendicular to the major axis, and also bisect the respective Cl–I–Cl angles. The double-prime superscript serves to distinguish these axes from the C_2 and C_2' axes. The C_2'' axes are contained by two extra planes of symmetry (they are the xz and yz planes), which are termed dihedral planes since they bisect the two C_2' axes. The dihedral planes contain the major axis, C_4.

Improper Axes of Symmetry, S_n

If rotation about an axis by 360°/n followed by reflexion through a plane perpendicular to the axis produces an equivalent configuration of a molecule, then the molecule contains an **improper axis of symmetry**. Such an axis is denoted by S_n, the associated symmetry operation having been described in the previous sentence. The C_3 axis of the PCl_5 molecule is also an S_3 axis. The operation of S_3 on PCl_5 causes the apical (*i.e.* out-of-plane) chlorine atoms to exchange places.

The operation of reflexion through a horizontal plane may be regarded as a special case of an improper axis of symmetry of order one: S_1. The rotation of a molecule around an axis by 360° produces an identical configuration ($C_1 = E$), and the reflexion in the horizontal plane is the only non-trivial part of the operations associated with the S_1 improper axis. This may be symbolized as:

$$S_1 = C_1 \times \sigma_h = E \times \sigma_h = \sigma_h \qquad (2.1)$$

If a molecule does not possess an improper axis of symmetry it is termed **dissymmetric** and cannot have a mirror image that is superposable on itself. Molecules in this class are termed **chiral**, *i.e.* they can show **optical activity** such that optical isomers deflect the plane of polarized light passing through their solutions in opposite directions.

The Inversion Centre or Centre of Symmetry, *i*

An **inversion centre** or **centre of symmetry** (denoted by *i*) is possessed by a molecule which has pairs of identical atoms that are diametrically opposed to each other about the centre. Any particular atom with the coordinates x, y, z must be partnered by an identical atom with the coordinates $-x, -y, -z$, if the molecule is to possess an inversion centre. That condition must apply to all atoms in the molecule which are off-centre. The ICl_4^- ion possesses an inversion centre, as does the dihydrogen molecule, H_2. The BF_3 molecule (trigonally planar) does not possess an inversion centre. The symmetry operation associated with an inversion centre is the inversion of the molecule, in which the diametrically opposed atoms in each such pair exchange places. The inversion centre is another special case of an improper axis in that the operations associated with the S_2 element are exactly those which produce the inversion of the atoms of a molecule containing *i*. Reference to Figure 2.5 shows the effects of carrying out the operation C_2 followed by σ_h, and that the fluorine atoms in each of the three diametrically opposed pairs in the octahedral SF_6 molecule have been exchanged with each other, all of which is what is understood to be the inversion of the molecule. This is symbolized as:

$$S_2 = C_2 \times \sigma_h = i \tag{2.2}$$

The superscript numbering of the fluorine atoms in Figure 2.5 is done to make clear the atomic movements which take place as a result of the application of the symmetry operations.

Figure 2.5 An illustration of the inversion of the SF_6 molecule by two paths

> **Worked Problem**
>
> **Q** Does the tetrahedral CH_4 molecule possess an inversion centre?
>
> **A** No, because the hydrogen atoms do not occur in diametrically opposite pairs. If the molecule were planar, then it would possess an inversion centre.

2.2   Point Groups and Character Tables

The symmetry elements which may be possessed by a molecule are defined above and exemplified. The next stage is to decide which, and how many, of these elements are possessed by particular molecules so that the molecules can be assigned to **point groups**. A point group consists of *all the elements of symmetry possessed by a molecule and which intersect at a point*. Such elements represent a group according to the rules to be outlined. The interactions of the properties of the elements of symmetry of a point group are summarized in a **character table**. The description of point groups and character tables is best carried out by means of examples. A simple example is the water molecule, which is a bent triatomic system.

2.2.1 Point Groups

The symmetry elements of the water molecule are easily detected. There is only one proper axis of symmetry, which is the one that bisects the bond angle and contains the oxygen atom. It is a C_2 axis and the associated operation of rotating the molecule about the axis by 180° results in the hydrogen atoms exchanging places with each other. The demonstration of the effectiveness of the operation is sufficient for the diagnosis of the presence of the element.

Figure 2.6 shows the molecule of water set up so that the C_2 axis is coincident with the z axis according to convention. The molecular plane is set up to be in the yz plane so that the x axis is perpendicular to the paper, as is the xz plane. There are two vertical planes of symmetry: the xz and the yz planes, and these are designated as σ_v and σ_v', respectively, the prime serving to distinguish between the two. The only other symmetry element possessed by the water molecule is the identity, E.

The four symmetry elements form a group, which may be demonstrated by introducing the appropriate rules. The rules are exemplified by considering the orbitals of the atoms present in the molecule. Such a consideration also develops the relevance of **group theory**, in that it leads

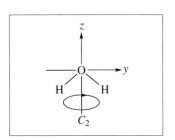

Figure 2.6 The water molecule set up in the *yz* plane with the *z* axis containing the C_2 axis

to an understanding of which of the atomic orbitals are permitted to combine to form molecular orbitals.

The electrons which are important for the bonding in the water molecule are those in the **valence shell** of the oxygen atom: $2s^2 2p^4$. It is essential to explore the character of the 2s and 2p orbitals, and this is done by deciding how each orbital **transforms** with respect to the operations associated with each of the symmetry elements possessed by the water molecule.

The Characters of the Representation of the 2s Orbital of the Oxygen Atom

The **character** of an orbital is symbolized by a number which expresses the result of any particular operation on its wave function. In the case of the 2s orbital of the oxygen atom which is spherically symmetrical, there is no change of sign of ψ with any of the four operations: E, C_2 (z), σ_v (xz) and σ_v' (yz). These results may be written down in the form:

	E	C_2	σ_v (xz)	σ_v' (yz)
2s	1	1	1	1

the '1's indicating that the 2s orbital is **symmetric** with respect to the individual operations. Such a collection of characters is termed a **representation**, this particular example being the **totally symmetric representation** (*i.e.* all the characters are '1's).

The Characters of the Representation of the $2p_x$ Orbital of the Oxygen Atom

The $2p_x$ orbital of the oxygen atom is symmetric with respect to the identity, but is **anti-symmetric** with respect to the C_2 operation. This is because of the spatial distribution of ψ values in the 2p orbitals with one positive lobe and one negative lobe. The operation, C_2, causes the positive and negative regions of the $2p_x$ orbital to exchange places with each other. This sign reversal is indicated as a character of -1 in the representation of the $2p_x$ orbital as far as the C_2 operation is concerned. The $2p_x$ orbital is unchanged by reflexion in the xz plane but suffers a sign reversal when reflected through the yz plane. The collection of the characters of the $2p_x$ orbital with respect to the four symmetry operations associated with the four elements of symmetry forms another representation of the group currently being constructed:

	E	C_2	$\sigma_v\ (xz)$	$\sigma_v'\ (yz)$
$2p_x$	1	−1	1	−1

The Characters of the Representation of the $2p_y$ Orbital of the Oxygen Atom

The $2p_y$ orbital of the oxygen atom changes sign when the C_2 operation is applied to it, and when it is reflected through the xz plane, but is symmetric with respect to the molecular plane, yz. The representation expressing the character of the $2p_y$ orbital is another member of the group:

	E	C_2	$\sigma_v\ (xz)$	$\sigma_v'\ (yz)$
$2p_y$	1	−1	−1	1

The Characters of the Representation of the $2p_z$ Orbital of the Oxygen Atom

The $2p_z$ orbital of the oxygen atom has exactly the same set of characters as the 2s orbital: it is another example of a totally symmetric representation.

	E	C_2	$\sigma_v\ (xz)$	$\sigma_v'\ (yz)$
$2p_z$	1	1	1	1

The Multiplication of Representations

At this stage it is important to use one of the rules of group theory, which states that *the product of any two representations of a group must also be a member of that group*. This rule may be used on the examples of the representations for the $2p_x$ and $2p_y$ orbitals as deduced above. The product of two representations is obtained by multiplying together the individual characters for each symmetry element of the group. The normal rules of arithmetic apply, so that the representation of the product of those of the two 2p orbitals under discussion is given by:

	E	C_2	$\sigma_v\ (xz)$	$\sigma_v'\ (yz)$
$2p_x$	1	-1	1	-1
$2p_y$	1	-1	-1	1
$2p_x \times 2p_y$	1	1	-1	-1

The new representation is, by the rules, a member of the group, and is also the only possible addition to the set of representations so far deduced for the molecular shape under consideration. There are no other combinations of 1 and -1 that would form a different representation.

Exercise

The above statement may be checked by trying other double products among the four representations.

The four representations may be collected together and be given symbols, the collection of characters forming what is known as a **character table**:

	E	C_2	σ_v	σ_v'
A_1	1	1	1	1
A_2	1	1	-1	-1
B_1	1	-1	1	-1
B_2	1	-1	-1	1

The symbols used for the representations are those proposed by Mulliken. The A representations are those which are symmetric with respect to the C_2 operation, and the Bs are antisymmetric to that operation. The subscript 1 indicates that a representation is symmetric with respect to the σ_v operation, the subscript 2 indicating antisymmetry to it. No other indications are required, since the characters in the σ_v' column are decided by another rule of group theory. This rule is: *the product of any two columns of a character table must also be a column in that table*. It may be seen that the product of the C_2 characters and those of σ_v give the contents of the σ_v'. The representations deduced above must be described as **irreducible representations**. This is because they cannot be simpler, but there are other representations (examples follow) which are **reducible** in that they are sums of irreducible ones. Character tables, in general, are a list of the irreducible representations of the particular group, and as in the ones shown in Appendix 1 they have an extra column which indicates the representations by which various orbitals transform.

In the example under discussion the 2s and $2p_z$ orbitals transform as a_1, the $2p_x$ orbital transforms as b_1 and the $2p_y$ orbital transforms as b_2. In this context 'transform' refers to the behaviour of the orbitals with respect to the symmetry operations associated with the symmetry elements of the particular group.

It should be noticed that lower case Mulliken symbols are used to indicate the irreducible representations of orbitals. The upper case Mulliken symbols are reserved for the description of the symmetry properties of electronic states.

Symmetry Properties of the Hydrogen 1s Orbitals in the Water Molecule; Group Orbitals

Neither of the 1s orbitals of the hydrogen atoms of the water molecule, taken separately, transform within the group of irreducible representations deduced for that molecule. The two 1s orbitals must be taken together as one or the other of two **group orbitals**. A more formal treatment of the group orbitals which two 1s orbitals may form is dealt with in Chapter 3.

Whenever there are two or more identical atoms linked to a central atom, their wave functions must be combined in such a way as to demonstrate their indistinguishability. This is achieved by making **linear combinations** of the wave functions of the ligand atoms. For the two 1s orbitals of the hydrogen atoms in the water molecule their wave functions may be combined to give:

The term **ligand** is used to describe any atom or group of atoms which is chemically bonded to a central atom, *i.e.* an atom at the coordinate origin. In the example under discussion the oxygen atom is at the centre of the point group and the two hydrogen atoms may be described as ligands.

$$h_1 = 1s_A + 1s_B \tag{2.3}$$

$$h_2 = 1s_A - 1s_B \tag{2.4}$$

where h_1 and h_2 are the wave functions of the two group orbitals, and $1s_A$ and $1s_B$ represent the 1s wave functions of the two hydrogen atoms, A and B. The two group orbitals are shown diagrammatically in Figure 2.7. By inspection, they may be shown to transform as a_1 and b_2 representations, respectively. The h_1 orbital behaves exactly like the 2s and $2p_z$ orbitals of the oxygen atom, and the h_2 orbital behaves exactly like the $2p_y$ orbital of the oxygen atom, with respect to the four symmetry operations of the point group.

h_1	E	C_2	$\sigma_v\,(xz)$	$\sigma_v'\,(yz)$
	1	1	1	1

h_2	E	C_2	$\sigma_v\,(xz)$	$\sigma_v'\,(yz)$
	1	−1	−1	1

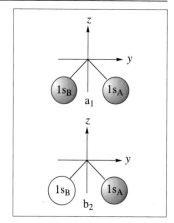

Figure 2.7 The (*top*) a$_1$ and (*bottom*) b$_2$ group orbitals formed from the 1s atomic orbitals of the two hydrogen atoms of the water molecule

2.2.2 Character Tables

It is necessary to label the various point groups to which molecules may belong. The labelling system adopted is to use a letter which is related to the major axis, and to use the value of n (the order of the major axis) together with a letter indicating the plane of symmetry (h, v or d) of highest importance for descriptive purposes, as subscripts. The system was suggested by Schönflies and the labels are known as **Schönflies symbols**. The method of deciding the point group of a molecule is that described below.

Assignment of Molecules to Point Groups

There are three shapes which are particularly important in chemistry and are easily recognized by the number of their faces, all of which consist of equilateral triangles. They are the **tetrahedron** (four faces), the **octahedron** (eight faces) and the **icosahedron** (20 faces). Figure 2.8 contains diagrams of these shapes and examples from chemistry.

The first two of the shapes are extremely common in chemistry, while the third shape is important in boron chemistry and many other **cluster** molecules (a cluster is defined as a molecule in which three or more identical atoms are bonded to each other) and ions. The three special shapes are associated with point groups and their character tables and are labelled, T_d, O_h and I_h, respectively. The point group to which a molecule belongs may be decided by the answers to four main questions:

1. Does the molecule belong to one of the special point groups, T_d, O_h or I_h? If it does, the point group has been identified.
2. If the molecule does not belong to one of the special groups, it is necessary to identify the major axis (or axes). The major axis is the one with the highest order (the highest value of n) and is designated C_n. In a case where there is more than one axis which could be classified as major (of equal values of n), it is conventional to regard the axis placed along the z axis as being the major one.

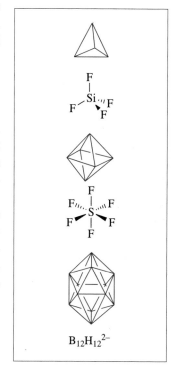

Figure 2.8 The 'special' shapes, the tetrahedron, the octahedron and the icosahedron with chemical examples of each: SiF$_4$, SF$_6$, [B$_{12}$H$_{12}$]$^{2-}$

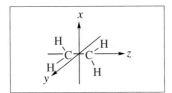

Figure 2.9 The structure of the ethene molecule showing its three C_2 axes

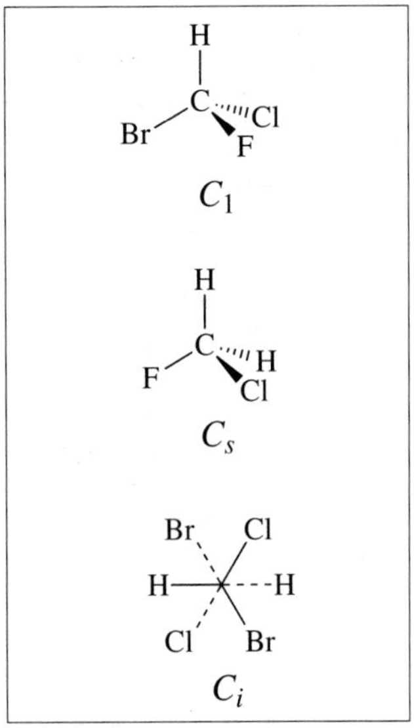

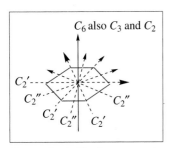

Figure 2.10 Examples of C_1, C_s and C_i molecules: CHBrClF, CH$_2$ClF and staggered C$_2$(HBrCl)$_2$ looking along the C–C axis

Figure 2.11 The benzene molecule showing the C_6 axis and the six C_2 axes at 90° to it

---

Worked Problems

Q The BF$_3$ molecule possesses a C_3 axis and three C_2 axes. Which is the major axis?

A The C_3 axis is the major axis and is normally positioned along the Cartesian z axis.

Q The C$_2$H$_4$ molecule has three C_2 axes. Which is the major axis?

A There is no clear major axis. It is normal to make the C_2 axis which contains the two carbon atoms coincident with the Cartesian z axis, as shown in Figure 2.9, to be the major axis.

In the trivial cases where there is only a C_1 axis, the point group of the molecule is C_1, unless there is either a plane of symmetry or an inversion centre present, indicating the point groups C_s or C_i, respectively. Examples of C_1, C_s and C_i molecules are shown in Figure 2.10.

The main question is: are there n C_2 axes perpendicular to the major axis, C_n? If there are, the molecule belongs to a D_n group, otherwise the molecule belongs to a C_n group. Figure 2.11 shows the benzene molecule, which belongs to a D_n group. In the benzene molecule the major axis is C_6, perpendicular to the molecular plane, and there are six C_2 axes perpendicular to the major axis. Benzene belongs to a D_n group, therefore. Although the ethene molecule, shown in Figure 2.9, does not possess a major axis, it does have three C_2 axes which are mutually perpendicular, and so belongs to a D_n group in which the z axis may be regarded as major, with the C_2 axes coincident with the x and y axes satisfying the qualification.

3. The third question applies only to D_n groups: is there a horizontal plane of symmetry present? If σ_h is present, the molecule belongs to the point group D_{nh}. As shown in Figure 2.11, the benzene molecule does possess a horizontal plane and belongs to the point group D_{6h}. If there is no σ_h but there are n σ_v present, the molecule belongs to the point group D_{nd}. The example shown in Figure 2.12 is the allene molecule, which has a C_2 axis (major) coincident with the CCC axis with two dihedral planes which contain that axis and both contain and bisect respectively the two CH$_2$ groups. If no σ_v is present, then the point group of the molecule is D_n. As shown in Figure 2.13, the staggered form of the ethane molecule has a C_3 major axis with three C_2 axes which are mutually perpendicular to it, but there are no planes of symmetry so the molecule belongs to the D_3 point group.

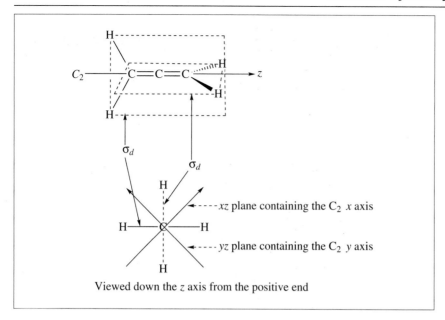

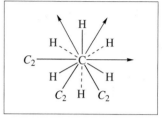

Figure 2.12 (Left) The allene molecule which has a C_2 axis (major) coincident with the CCC axis with two dihedral planes which contain that axis and both contain and bisect respectively the two CH_2 groups

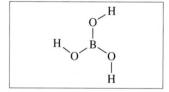

Figure 2.13 The staggered form of the ethane molecule, looking down the C_3 major axis, and showing the three C_2 axes perpendicular to it

4. The fourth question applies only to C_n groups: is there a σ_h present? If there is, the molecule belongs to the point group C_{nh}. There are very few examples of such molecules, and Figure 2.14 shows the planar form of $B(OH)_3$ which has a C_3 major axis perpendicular to the molecular plane, but there are no other planes of symmetry. It belongs, therefore, to the point group C_{3h}.

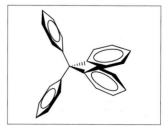

Figure 2.14 The planar $B(OH)_3$ molecule; its C_3 axis passes through the boron nucleus and is at right angles to the molecular plane

If there is no σ_h but there are n σ_v present, the molecule belongs to the point group C_{nv}. Examples of these molecules are common in chemistry; two are shown in Figure 2.15.

The trigonally pyramidal ammonia molecule has a C_3 axis with three vertical planes passing through it and belongs to the point group C_{3v}. If no σ_v is present but there is an S_{2n} improper axis present, the molecule belongs to the S_{2n} point group. Very few examples of S_{2n} molecules exist. One is tetraphenylmethane, shown in Figure 2.16.

An S_4 operation causes pairs of phenyl groups to interchange with one another to produce an equivalent structure of the molecule, so the molecule belongs to the point group S_4. If no S_{2n} is present, the point group of the molecule is C_n. At this point, molecules are severely lacking in symmetry elements. There are molecules such as CFClBrI, which belongs to C_1 (this is equivalent to the identity, E), and H_2O_2 with its non-planar arrangement of OH groups, as shown in Figure 2.17, which belongs to C_2. In Figure 2.17, one of the oxygen atoms is placed on top of the other one, so the view of the molecule is down the O–O axis, which is itself not an axis of symmetry.

The procedure described will identify the point groups of the majority

Figure 2.15 Examples of molecules belonging to C_{nv} groups: staggered Cl_3CCH_3 and $XeOF_4$

Figure 2.16 The structure of tetraphenylmethane; the view down any phenyl–central carbon direction is that of a propeller

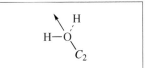

Figure 2.17 The structure of hydrogen peroxide, which belongs to the C_2 point group

of molecules. It may be written in the form of a flow sheet, which is shown in Figure 2.18 and allows the assignment of molecules to point groups. The most frequently used character tables are placed in Appendix 1.

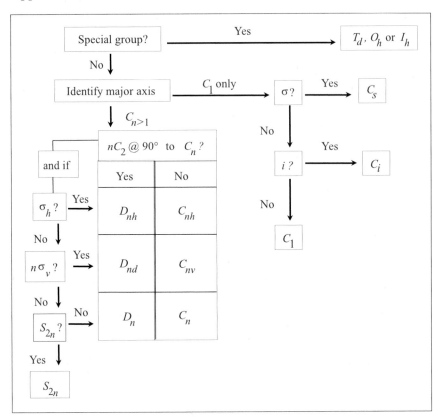

Figure 2.18 A flow chart for the assignment of molecules, groups of ligands or orbitals to point groups

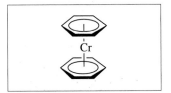

Figure 2.19 The structure of $Cr(C_6H_6)_2$

Worked Problems

Q In addition to the identity element, what symmetry elements are possessed by the molecules: (a) C_6H_5Br; (b) *cis*-$Pt(NH_3)_2Cl_2$ and *trans*-$Pt(NH_3)_2Cl_2$ (both planar structures; ignore the dispositions of the hydrogen atoms); (c) $AuBr_4^-$ (planar); (d) $(C_6H_6)_2Cr$, bis(benzene)chromium, a sandwich compound with eclipsed benzene rings, see Figure 2.19; (e) OCS (linear).

A (a) C_6H_5Br has only one axis of symmetry, the C_2 axis, which contains the line containing the C–Br bond. There are two vertical planes of symmetry: the molecular plane and the plane at 90° to that plane and which contains the C–Br bond. They both contain the C_2 axis.

(b) *cis*-$Pt(NH_3)_2Cl_2$ has the same symmetry elements as does the C_6H_5Br molecule. The C_2 axis bisects the Cl–Pt–Cl angle and is contained within the two vertical planes: the molecular plane and the plane at 90° to that plane and which both contain the Pt atom. *trans*-$Pt(NH_3)_2Cl_2$ has three mutually perpendicular C_2 axes. The axis passing through the Pt atom and which is perpendicular to the molecular plane is regarded as the major axis, so the molecular plane is to be regarded as horizontal. The other two C_2 axes, one passing through the Cl–Pt–Cl line and the other passing through the N–Pt–N line, are perpendicular to the major axis. There is an inversion centre at the centre of the Pt nucleus.

(c) $AuBr_4^-$ is a planar ion. There is an inversion centre at the Au atom. The major axis is C_4, passing through the Au atom and perpendicular to the molecular plane which is therefore a horizontal plane of symmetry. Coincident with the major axis is a C_2 axis. There are two C_2 axes of symmetry passing through the two Br–Au–Br directions and two more C_2 axes of symmetry which bisect the molecule, but only contain the Au atom. All four of the C_2 axes are perpendicular to the major axis. There are two vertical planes of symmetry which contain the major axis and each contains one of the linear Br–Au–Br groups. Two more planes of symmetry also contain the major axis and the Au atom, but bisect the molecule and bisect the two C_2 axes of symmetry which contain the two linear Br–Au–Br groups. These are dihedral planes. There is an S_4 improper axis of symmetry coincident with the major axis, although this does not seem relevant to the planar ion.

(d) $(C_6H_6)_2Cr$ has a C_6 major axis passing through the centres of the benzene rings and the Cr atom. Coincident with the major axis is a C_2 axis, a C_3 axis, an S_3 axis and an S_6 axis. There is a horizontal plane parallel to the planes of the benzene rings and which contains the Cr atom. There are six C_2 axes perpendicular to the major axis and which contain the Cr atom; they are parallel to the planes of the two benzene rings (they are the same as those shown in Figure 2.11 for the benzene molecule). There is an inversion centre at the Cr atom (because the two rings are in an eclipsed conformation). There are three vertical and three dihedral planes.

(e) OCS is linear; the molecular axis is the major axis of symmetry with an order of infinity, C_∞. There are an infinite number of vertical planes of symmetry which pass through the molecule and which contain the major axis.

Q To which point groups do the molecules of the previous question belong?

A (a) C_{2v}, because there is only one axis of symmetry and there are two vertical planes of symmetry.
(b) *cis*-Pt(NH$_3$)$_2$Cl$_2$ has the same symmetry properties as does C$_6$H$_5$Br and so belongs to the C_{2v} point group. *trans*-Pt(NH$_3$)$_2$Cl$_2$ passes the $nC_2 \perp C_n$ test, with $n = 2$, and there is a molecular horizontal plane, so the point group is D_{2h}.
(c) AuBr$_4^-$ has a major axis, C_4, and passes the $nC_2 \perp C_n$ test, with $n = 4$, and there is a molecular horizontal plane, so the point group is D_{4h}.
(d) (C$_6$H$_6$)$_2$Cr has the same symmetry properties as does the benzene molecule and belongs to the D_{6h} point group.
(e) OCS belongs to the $C_{\infty h}$ point group.

Q To which point group does a soccer ball with the C$_{60}$ fullerene (soccerene) pattern belong?

A This is possibly difficult to visualize and the shape is not obviously one of the special cases. The fullerene structure can be produced by taking an icosahedron (Figure 2.8) and slicing the tops of all the vertices in a regular fashion. The slicing operation produces 12 pentagons and also converts the 20 triangular faces of the icosahedron into hexagons. The football belongs to the icosahedral point group, I_h.

Q To which point group does a tennis ball belong?

A Tennis balls have a particular seam. A diagram of the ball and its seam is shown in Figure 2.20, which indicates the position of one of the C_2 axes and its relationship to the seam. There are two other C_2 axes which are perpendicular to the other axis, but there is no horizontal plane. The ball belongs to the D_{2d} point group.

Q To which point group does a rugby football belong, assuming it to be free of patterns?

A There is a C_∞ axis passing through the ball along its largest dimension and an infinite number of vertical planes containing that axis. There are also an infinite number of C_2 axes that are perpendicular to the major axis and which are contained by the horizontal plane. This passes the $nC_2 \perp C_n$ test with $n = \infty$, so the ball belongs to the $D_{\infty h}$ point group.

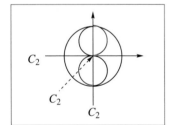

Figure 2.20 A diagram of a tennis ball showing the seam and the positions of the three C_2 axes. The C_2 operation perpendicular to the paper causes the interchanges of the two halves of the red curve (representing the seam on the underside of the ball) and of the two halves of the black curve (representing the seam on the topside of the ball). The other two C_2 operations cause the interchanges of black and red sections of the seam.

Summary of Key Points

1. The seven different elements of symmetry were described and their associated symmetry operations exemplified.

2. Point groups were defined as consisting of *all the elements of symmetry possessed by a molecule and which intersect at a point.* Such elements represent a group according to the rules by which they interact.

3. The interactions of the properties of the elements of symmetry of a point group were summarized in character tables.

4. A method of assigning molecules to particular point groups was outlined.

Problems

2.1. Allocate the following species to their appropriate point group (you may wish to postpone this exercise until you have read the sections about molecular shapes in Chapters 5 and 6):

(a) N_2O
(b) $HgCl_2(g)$
(c) N_3^-
(d) $AuCl_2^-$
(e) O_3
(f) ClO_2

(g) F_2O
(h) $ClIBr^-$
(i) S_3^{2-}
(j) H_2Se
(k) $HOCl$
(l) $OPCl_3$

(m) NO_2
(n) $SiBrClFI$
(o) IF_5
(p) PCl_3
(q) $ONCl$
(r) PO_4^{3-}

Further Reading

F. A. Cotton, *Chemical Applications of Group Theory*, 3rd edn., Interscience, New York, 1990. This book is the oldest, best and the most expensive on the topic.

3

Covalent Bonding I: the Dihydrogen Molecule-ion, H_2^+, and the Dihydrogen Molecule

This chapter consists of the application of the symmetry concepts of Chapter 2 to the construction of molecular orbitals for a range of diatomic molecules. The principles of molecular orbital theory are developed in the discussion of the bonding of the simplest molecular species, the one-electron dihydrogen molecule-ion, H_2^+, and the simplest molecule, the two-electron dihydrogen molecule. Valence bond theory is introduced and compared with molecular orbital theory. The photoelectron spectrum of the dihydrogen molecule is described and interpreted.

Aims

By the end of this chapter you should understand:

- The difference between covalent and ionic bonding
- The use of linear combinations of atomic orbitals (LCAO) to produce molecular orbitals
- The application of group theory to define the orbitals of the dihydrogen molecule
- The normalization of molecular orbitals
- The overlap integral
- The relative energies of atomic, bonding and anti-bonding orbitals
- The nature of a single covalent bond
- The basis of the valence bond theory
- The energetics of the dihydrogen molecule-ion, H_2^+, and the dihydrogen molecule
- The photoelectron spectrum of the dihydrogen molecule

3.1   Ionic, Covalent and Coordinate (or Dative) Bonding

There are two extreme cases of chemical bonding: ionic and covalent. **Ionic bonding** occurs when there is a complete transfer of one electron from one atom to another to form **ions** as in the equation:

$$Na(1s^22s^22p^63s^1) + F(1s^22s^22p^5) \rightarrow Na^+(1s^22s^22p^6) + F^-(1s^22s^22p^6) \quad (3.1)$$

In this case the electropositive sodium atom loses its 3s electron, which is then transferred to the 2p orbital of the electronegative fluorine atom to produce the Na^+F^- ion-pair. Ionic bonding is the subject of Chapter 7. When there is little or no difference in the electronegativity coefficients of the combining atoms, **covalent bonds** are possible in which two or more electrons are *shared* between the two atoms. Covalency is the main subject of this chapter.

The **coordinate bond**, also known as the **dative bond**, is an electron-pair bond between two atoms or groups of atoms in which both electrons are supplied by one atom. The atom supplying the electron pair is called the **donor**, the receiving atom being the **acceptor**. Although the coordinate bond is extremely important in the description of compounds of the transition elements, it is given minimal treatment here. The classic example of coordinate bond formation in main group chemistry is the formation of an **adduct** (*i.e.* an **addition compound**) from boron trifluoride and ammonia. The ammonia molecule possesses a non-bonding (lone) pair of electrons, which is used to form a coordinate bond with the boron trifluoride, which has a vacant p-type orbital:

$$H_3N: + BF_3 \rightarrow H_3N:\rightarrow BF_3 \quad (3.2)$$

The resulting adduct has the same number of valence electrons as the ethane molecule, C_2H_6, and has the same structure with the two parts of the molecule having rotational freedom around the N→B coordinate bond, or covalent C–C bond in ethane. Coordinate or dative bonds are usually drawn in molecular structures as arrows to represent the direction of the donation process as in: **donor atom→acceptor atom**

Reaction 3.2 is a prime example of the use of the terms **Lewis acid** and **Lewis base**. G. N. Lewis suggested the usage such that a donor of an electron pair is a base and the acceptor molecule is an acid. The classical **Brønsted** acid and base pair, H^+(aqueous) and OH^-(aqueous), are encompassed by the Lewis definitions as they combine to give water, the hydroxide ion supplying both electrons.

3.2   Covalent Bonding in H_2^+ and H_2

A covalent bond occurs when two atoms share two or more electrons. More specifically, in the context of molecular orbital theory, a single covalent bond between two atoms occurs when two electrons (one from each of the atoms) occupies a bonding molecular orbital. Other terms

The meanings of the terms **bonding**, **non-bonding** and **anti-bonding orbitals** are developed in stages in this chapter.

The **three-body problem** in physics is that the independent motion of three bodies, their velocities and positions, cannot be described by a sufficient number of equations so that an analytical solution may be achieved.

used in molecular orbital theory are non-bonding and anti-bonding orbitals.

The H_2^+ molecule-ion, which consists of two protons and one electron, represents an even simpler case of a covalent bond, in which only one electron is shared between the two nuclei. Even so, it represents a quantum mechanical **three-body problem**, which means that solutions of the wave equation must be obtained by **iterative methods**. The molecular orbitals derived from the combination of two 1s atomic orbitals serve to describe the electronic configurations of the four species H_2^+, H_2, He_2^+ and He_2.

3.2.1 The Construction of Molecular Orbitals

The basic concept of **molecular orbital theory** is that molecular orbitals may be constructed from a set of contributing atomic orbitals such that the molecular wave functions consist of **linear combinations of atomic orbitals (LCAO)**. In the case of the combination of two 1s hydrogen atomic orbitals to give two molecular orbitals, the two linear combinations are written below so that atomic wave functions are represented by ψ and **molecular wave functions by ϕ**:

Equations 3.3 and 3.4 seem to imply that both atomic wave functions ψ_A and ψ_B contribute fully (100%) to both molecular wave functions ϕ_1 and ϕ_2. The impossibility of this is taken care of in the section on **normalization**. For the purposes of this section, normalization may be ignored as it does not alter the symmetry properties of the molecular wave functions.

$$\phi_1 = \psi_A + \psi_B \tag{3.3}$$

$$\phi_2 = \psi_A - \psi_B \tag{3.4}$$

where ψ_A and ψ_B are the two hydrogen atomic 1s wave functions of atoms A and B, respectively. The two linear combinations may be represented by diagrams in which the envelopes of the two orbitals are used as shown in Figure 3.1. The two 1s orbitals as they are represented in equations 3.3 and 3.4 are shown overlapping in two ways.

Where the two atomic orbitals have the *same signs* for ψ there is an *increase in probability* of finding an electron in the internuclear region because the two atomic orbitals both contribute to a build-up of ψ value. The orbital so produced is called a **bonding molecular orbital**. If the two 1s orbitals have *different signs* for ψ, the internuclear region has a *virtually zero probability* for finding the electron because the two wave functions cancel out, and the molecular orbital produced is described as anti-bonding. This is shown in a more mathematical way later in this section. The formal way in which the forms of the two molecular orbitals are decided is by using group theory.

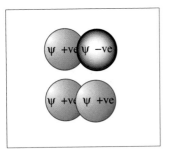

Figure 3.1 The two ways in which 1s orbitals from two hydrogen atoms are permitted to overlap. In the top diagram, one 1s wave function is positive and the other negative; in the lower diagram, both 1s wave functions are positive

Group Theory Applied to the Molecular Orbital Formation in H_2^+ and H_2

The H_2^+ and H_2 molecules belong to the $D_{\infty h}$ point group. The two 1s atomic orbitals, individually, do not transform within the $D_{\infty h}$ point group, but together their symmetry representation, *i.e.* the characters with respect to the various symmetry operations of the $D_{\infty h}$ point group, may be elucidated. This is done by considering each of the symmetry elements of the $D_{\infty h}$ group in turn, and writing down under each element *the number of orbitals unaffected by the associated symmetry operation*:

	E	C_∞^ϕ	σ_v	i	S_∞^ϕ	C_2
$\psi_A + \psi_B$	2	2	2	0	0	0

Box 3.1 Formal Determination of Characters

The 'rule-of-thumb' method for determining the character of a group of orbitals with respect to any particular symmetry operation, *i.e.* by writing down the number of orbitals which are unaffected by the operation, is extremely useful. The underlying logic of the rule is explained here in terms of the two 1s orbitals which participate in the formation of the covalent bonding of the dihydrogen molecule. There are two vectors $\mathbf{v}_A$ and $\mathbf{v}_B$ which can be considered to represent the displacement of the respective hydrogen atoms, A and B, from the centre of the molecule, the latter being the Cartesian zero on which the point group $D_{\infty h}$ is based. These vectors may or may not be altered by carrying out a symmetry operation on the molecule or the two orbitals concerned. If a C_∞ operation is carried out (rotation about the molecular axis by any angle) the two vectors are unchanged. This can be expressed mathematically by the equations:

$$\mathbf{v}_A = \mathbf{v}_A + 0\mathbf{v}_B \tag{3.5}$$

$$\mathbf{v}_B = 0\mathbf{v}_A + \mathbf{v}_B \tag{3.6}$$

The corresponding transformation matrix is:

$$\begin{bmatrix} 1 & 0 \\ 0 & 1 \end{bmatrix}$$

which has a *trace* (*i.e.* the sum of the numbers along the diagonal from top-left to bottom-right) of 2. Thus, 2 is the character of the orbitals with respect to the C_∞ operation. The same number corresponds to the number of orbitals unaffected by the operation.

The matrix representing the effects of the i operation is similarly generated from the equations:

$$\mathbf{v}_A = 0\mathbf{v}_A + \mathbf{v}_B \tag{3.7}$$

$$\mathbf{v}_B = \mathbf{v}_A + 0\mathbf{v}_B \tag{3.8}$$

$$\begin{bmatrix} 0 & 1 \\ 1 & 0 \end{bmatrix}$$

which has a trace of zero, again coinciding with the number of orbitals that are unaffected by the operation, in this case zero, both having swapped places with each other.

The short-cut method of determining a character of a set of orbitals by deciding on the number of them unaffected by a symmetry operation is usually straightforward, although complications can arise with some point groups.

The two 1s orbitals are unaffected by the E (identity) operation, and hence the number 2 is written down in the representation. Rotation by any angle, ϕ, around the C_∞ axis does not affect the orbitals; hence the second 2 appears as the character of the two 1s orbitals. The third 2 appears because the two orbitals are unaffected by reflexion in any of the infinite number of vertical planes which contain the molecular axis. The operation of inversion affects both orbitals in that they exchange places with each other, and so a zero is written down in the i column. Likewise an S_∞ operation causes the orbitals to exchange places and a zero is written in that column. There are an infinite number of C_2 axes passing through the inversion centre and these are perpendicular to the molecular axis. The associated operation of rotation through 180° around any C_2 axis causes the 1s orbitals to exchange places with each other, so there is a final zero to be placed in the representation.

The sequence of numbers arrived at constitutes the representation of the two 1s orbitals with respect to $D_{\infty h}$ symmetry. Such a combination of numbers is not to be found in the $D_{\infty h}$ character table: it is an example of a **reducible representation**. Its reduction to a sum of **irreducible representations** is, in this instance, a matter of realizing that the sum of the σ_g^+ and σ_u^+ characters is the representation of the two 1s orbitals:

	E	C_∞^ϕ	σ_v	i	S_∞^ϕ	C_2
σ_g^+	1	1	1	1	1	1
σ_u^+	1	1	1	-1	-1	-1
$\sigma_g^+ + \sigma_u^+$	2	2	2	0	0	0

There is a formal method of converting a reducible representation to a sum of irreducible representations, but commonly the outcome is either obvious or may be arrived at by a few guesses. The formal method is described in Appendix 2.

Lower case letters are used for the symbols representing the symmetry properties of orbitals. Greek letters are used to symbolize the irreducible representations of the $D_{\infty h}$ point group, and have either g or u subscripts and either + or – signs as superscripts. The g and u subscripts refer to the character in the i column: g (coming from the German word *gerade*, meaning *even*) indicating symmetrical behaviour and u (*ungerade*, meaning *odd*) indicating antisymmetrical behaviour with respect to inversion. The + and – signs refer respectively to symmetry and antisymmetry with respect to reflexion in one of the vertical planes. The '2's in the E and some of the other columns are indications of **doubly degenerate** representations (some representations in other point groups are triply degenerate, and this is indicated by a '3' in the E column).

By referring to the diagrams in Figure 3.1 it may be seen that the orbital ϕ_1 transforms as σ_g^+, and that the orbital ϕ_2 transforms as σ_u^+. That two molecular orbitals are produced from the two atomic orbitals is an important part of molecular orbital theory: a **law of conservation of orbital numbers**. The two molecular orbitals differ in energy, both from each other and from the energy of the atomic level. To understand how this arises it is essential to consider the **normalization** of the orbitals.

Normalization is the procedure of arranging for the integral over all space of the square of the orbital wave function to be unity, as described in Section 1.3. For molecular orbitals this is expressed by the equation:

$$\int_0^\infty \phi^2 d\tau = 1 \tag{3.9}$$

where $d\tau$ is a volume element (equal to $dx.dy.dz$). The probability of unity expresses the certainty of finding an electron in the orbital. It must be that, by transforming atomic orbitals into molecular ones, no loss, or gain, in electron probability should occur. For equation 3.9 to be valid, **normalization factors**, N, must be introduced into equations 3.3 and 3.4, which then become:

$$\phi_1 = N_1(\psi_A + \psi_B) \tag{3.10}$$

$$\phi_2 = N_2(\psi_A - \psi_B) \tag{3.11}$$

where N_1 and N_2 are the appropriate normalization factors for the two molecular orbitals.

To determine the value of N_1 in equation 3.10 the expression for ϕ_1 must be placed into equation 3.9, giving:

$$N_1^2 \int_0^\infty (\psi_A + \psi_B)^2 d\tau = 1 \tag{3.12}$$

and expanding the square-term in the integral gives:

$$N_1^2 \int_0^\infty (\psi_A^2 + 2\psi_A \psi_B + \psi_B^2) d\tau = 1 \tag{3.13}$$

which may be written as three separate integrals:

$$N_1^2 \left(\int_0^\infty \psi_A^2 d\tau + 2\int_0^\infty \psi_A \psi_B d\tau + \int_0^\infty \psi_B^2 d\tau \right) = 1 \tag{3.14}$$

The assumption that the atomic orbital wave functions are separately normalized leads to the conclusion that the first and third integrals in equation 3.14 are both equal to 1. The second integral is known as the **overlap integral**, symbolized by S. This allows equation 3.14 to be simplified to:

$$N_1^2(1 + 2S + 1) = 1 \tag{3.15}$$

which gives a value for N_1 of:

$$N_1 = 1/(2 + 2S)^{1/2} \tag{3.16}$$

Worked Problem

Q In a similar manner to that used for the molecular orbital ϕ_1, carry out the normalization of ϕ_2, *i.e.* calculate a value for N_2 in equation 3.11.

A To determine the value of N_2 in equation 3.11 the expression for ϕ_2 must be placed into equation 3.9, giving:

$$N_2^2 \int_0^\infty (\psi_A - \psi_B)^2 d\tau = 1 \tag{3.17}$$

and expanding the square-term in the integral gives:

$$N_2^2 \int_0^\infty (\psi_A^2 - 2\psi_A \psi_B + \psi_B^2) d\tau = 1 \tag{3.18}$$

which may be written as three separate integrals:

$$N_2^2 \left(\int_0^\infty \psi_A^2 d\tau - 2\int_0^\infty \psi_A \psi_B d\tau + \int_0^\infty \psi_B^2 d\tau \right) = 1 \qquad (3.19)$$

The assumption that the atomic orbital wave functions are separately normalized leads to the conclusion that the first and third integrals in equation 3.19 are both equal to 1. The second integral is (as in equation 3.15) known as the **overlap integral**, symbolized by S. This allows equation 3.19 to be simplified to:

$$N_2^2(1 - 2S + 1) = 1 \qquad (3.20)$$

which gives a value for N_2 of:

$$N_2 = 1/(2 - 2S)^{1/2} \qquad (3.21)$$

To reinforce the meanings of diagrams such as those shown in Figure 3.1, it is helpful to plot the atomic and molecular wave functions along an axis which coincides with the molecular axis of H_2. For this exercise the wave function for the 1s orbital of the hydrogen atom may be written as:

$$\psi = Ne^{-r/a_0} \qquad (3.22)$$

where N is the normalizing factor for the atomic orbital. This wave function may be plotted for the two hydrogen atoms, A and B, with their respective origins separated by the known bond length of the dihydrogen molecule, 74 pm. So that the plots will represent a 'slice' through the wave functions along the molecular axis and to take into account the negative values of the distance from the nucleus at the origin, equation 3.22 may be modified to read:

$$\psi_A = Ne^{-|z|/a_0} \qquad (3.23)$$

which now contains the absolute value of the z coordinate (*i.e.* $|z|$, the value independent of sign), as a representation of the distance from the nucleus at the origin, that of atom A. Placing the second nucleus B at a distance R, equal to the bond length, further along the z axis in a positive direction is accomplished by modifying equation 3.23 to read:

$$\psi_B = Ne^{-|z-R|/a_0} \tag{3.24}$$

Plots of both equations 3.23 and 3.24 are shown in Figure 3.2, the value of N being taken to be 1 rather than the term $\pi^{1/2}a_0^{-3/2}$. The normalization factor can be ignored, for the purposes of Figure 3.2, since it is a scalar multiplier for all the orbital wave functions under consideration. Figure 3.2 also shows equation 3.24 plotted with negative values of the wave function (*i.e* $-\psi_B$), as is consistent with those it has in the formation of the anti-bonding wave function.

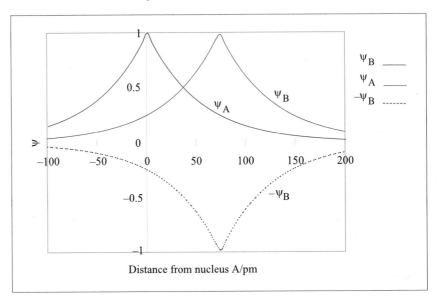

Figure 3.2 Plots of equations 3.23 and 3.24; equation 3.24 is also plotted with negative values

The equations for the bonding and anti-bonding wave functions used in Figure 3.3 are:

$$\phi_1 = \phi_{bonding} = \psi_A + \psi_B \tag{3.25}$$

$$\phi_2 = \phi_{anti\text{-}bonding} = \psi_A - \psi_B \tag{3.26}$$

The plots of equations 3.25 and 3.26 are shown in Figure 3.3, again considering the normalization factors to be 1.0. In the two plots shown in Figure 3.3 there is an *enhancement* of the ϕ values in the internuclear region of the bonding combination, but an *enfeeblement* in that region for the anti-bonding combination.

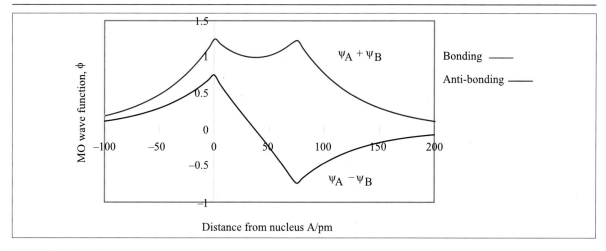

Worked Problem

Q In a diagram similar to that of Figure 3.3, plot the *squares* of the values of ϕ_1 and ϕ_2 to show the distribution of electron probability density along the molecular axis for H_2 and H_2^+. The plots are best produced by using a spreadsheet.

A The necessary equations that represent the squares of ϕ_1 and ϕ_2 are:

$$\phi_1^2 = \left(e^{-|z|/a_0} + e^{-|z-R|/a_0}\right)^2 \tag{3.27}$$

and

$$\phi_2^2 = \left(e^{-|z|/a_0} - e^{-|z-R|/a_0}\right)^2 \tag{3.28}$$

Your plots should be similar to those shown in Figure 3.4. That figure shows that there is a building-up of electron probability in the internuclear region for the bonding orbital, and that there is *zero probability* in the bond centre for the anti-bonding orbital.

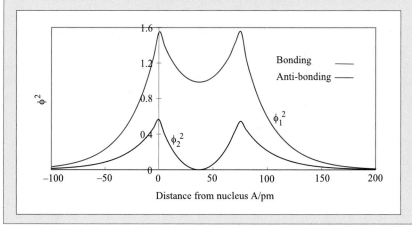

Figure 3.4 Plots of equations 3.27 and 3.28 for the bonding and anti-bonding wave functions of the dihydrogen molecule

The equations for the bonding and anti-bonding molecular orbital wave functions for the linear combinations of the two 1s atomic orbitals of the two hydrogen atoms may now be written in the forms:

$$\phi_1 = \phi_{\text{bonding}} = \left[\frac{1}{\left(2+2S\right)^{\frac{1}{2}}}\right]\left(\psi_A + \psi_B\right) \tag{3.29}$$

and

$$\phi_2 = \phi_{\text{anti-bonding}} = \left[\frac{1}{\left(2-2S\right)^{\frac{1}{2}}}\right]\left(\psi_A - \psi_B\right) \tag{3.30}$$

3.2.2 The Energies of Electrons in Molecular Orbitals

The next stage in the full description of the molecular orbitals is to calculate their energies, and to show that electrons in bonding molecular orbitals have a lower energy than when they occupy one or other of the constituent atomic orbitals and that electrons in anti-bonding orbitals are at a higher energy. This is done by considering the Schrödinger equation for molecular wave functions:

$$H\phi = E\phi \tag{3.31}$$

An example of the importance of the order of operations and what they operate on is to consider the operator: differentiate twice by x, $i.e.$ d^2/dx^2. Taking the function $3x^4$, the product equivalent to the order: function times the result of operating on the function is: $3x^4 \times d^2(3x^4)/dx^2$ $= 3x^4 \times 36x^2 = 108x^6$, but if the operator d^2/dx^2 is allowed to operate on the square of $3x^4$ the result is different: $d^2(9x^8)/dx^2 = 504x^6$.

If both sides are pre-multiplied by ϕ (this is essential for equations containing operators such as H) this gives:

$$\phi H\phi = \phi E\phi = E\phi^2 \tag{3.32}$$

there being no difference between $E\phi^2$ and $\phi E\phi$ since E is not an operator. Equation 3.32 may be integrated over all space to give:

$$\int_0^\infty \phi H\phi \, d\tau = E\int_0^\infty \phi^2 d\tau = E \tag{3.33}$$

since the integral on the right-hand side is equal to unity for normalized orbitals. Equation 3.33 may be used to calculate the energy of an electron in the molecular orbital ϕ_1, by substituting its value from equation 3.10:

$$E(\phi_1) = N_1^2 \int_0^\infty (\psi_A + \psi_B)H(\psi_A + \psi_B)d\tau$$

$$= N_1^2\left(\int_0^\infty \psi_A H\psi_A d\tau + \int_0^\infty \psi_B H\psi_B d\tau + 2\int_0^\infty \psi_A H\psi_B d\tau\right)$$

$$= N_1^2(\alpha + \alpha + 2\beta) = 2N_1^2(\alpha + \beta) \tag{3.34}$$

The first two integrals are entirely concerned with atomic orbitals, ψ_A and ψ_B, respectively, and have identical values (since they refer to identical orbitals) which are put equal to α, which is a quantity known as the Coulomb integral. In essence it is the energy of an electron in the 1s orbital of the hydrogen atom and equal to the ionization energy of that atom. The third integral is the sum of two identical integrals (again because ψ_A and ψ_B are identical) and is put equal to 2β, where β is called the resonance integral. β represents the *extra energy* gained by an electron, over that which it possesses in any case by being in the 1s atomic orbital of the hydrogen atom, when it occupies the molecular orbital ϕ_1. Because the electron is more stable in ϕ_1 than it is in ψ_A or ψ_B, ϕ_1 is called a **bonding molecular orbital**. Occupancy of ϕ_1 by one or two electrons leads to the stabilization of the system.

If the value of N_1 from equation 3.16 is substituted into equation 3.34 the expression for the energy of the bonding orbital becomes:

$$E(\phi_1) = \frac{\alpha + \beta}{1 + S} \tag{3.35}$$

Worked Problem

Q Carry out a similar treatment of the orbital ϕ_2 to produce an equation expressing the energy of an electron occupying the orbital.

A Take equation 3.11 and put the expression for ϕ_2 into equation 3.33 and carry out the same sequence of multiplications and substitutions:

$$E(\phi_1) = N_2^2 \int_0^\infty (\psi_A - \psi_B)H(\psi_A - \psi_B)\,d\tau$$

$$= N_2^2 \left(\int_0^\infty \psi_A H \psi_A\,d\tau + \int_0^\infty \psi_B H \psi_B\,d\tau - 2\int_0^\infty \psi_A H \psi_B\,d\tau \right)$$

$$= N_2^2(\alpha + \alpha - 2\beta) = 2N_2^2(\alpha - \beta) \tag{3.36}$$

Then substitute the value for N_2 from equation 3.21 to give the final equation for the energy of an electron in the anti-bonding orbital:

$$E(\phi_2) = \frac{\alpha - \beta}{1 - S} \tag{3.37}$$

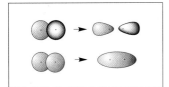

Figure 3.5 Orbital envelope diagrams showing the formation of the bonding and anti-bonding molecular orbitals when two 1s atomic orbitals overlap

Equation 3.37 shows that an electron in ϕ_2 has a higher energy than in ϕ_1. The electron energy in the anti-bonding orbital is also higher than if it were to occupy ψ_A (or ψ_B). The orbital ϕ_2 is therefore called an **anti-bonding** molecular orbital. Electrons in anti-bonding orbitals are less stable than in the atomic orbitals from which the molecular orbital was constructed. Such anti-bonding electrons contribute towards a weakening of the bonding of the molecule, or sometimes cause the complete dissociation of the molecule.

Using orbital envelope diagrams, the formation of bonding and anti-bonding orbitals from two 1s atomic orbitals is shown in Figure 3.5. In the bonding orbital the electron probability is concentrated in the internuclear region and is understandably an electrostatically attractive arrangement of the atomic particles. In the anti-bonding orbital the main electron probability is concentrated in regions other than the internuclear one and is understandably an electrostatically repulsive arrangement. The two nuclei are face-to-face with no negative electrons between them to provide electrostatic stability and cohesion.

Figure 3.6 shows the relative energies of the atomic orbitals, ψ_A and ψ_B, and the molecular orbitals, ϕ_1 and ϕ_2, which are involved in the production of H_2^+ and H_2. Both α and β are negative quantities on an energy scale with the ionization limit as the reference zero. The electronic configuration of the H_2^+ molecule-ion may be written as ϕ_1^1, or in symmetry symbols as $(\sigma_g^+)^1$. That of the dihydrogen molecule is ϕ_1^2, or $(\sigma_g^+)^2$, provided that the stabilization of ϕ_1 with respect to the atomic state is sufficiently large to force the electrons to pair up in the bonding orbital. The configuration, $\phi_1^1\phi_2^1$, produced by the absorption of a suitable quantum of energy, would lead to dissociation into separate hydrogen atoms. The occupation of the bonding orbital by a pair of electrons is the simplest example of a **single covalent bond**.

Figure 3.6 The molecular orbital diagram for the combination of two identical 1s atomic orbitals

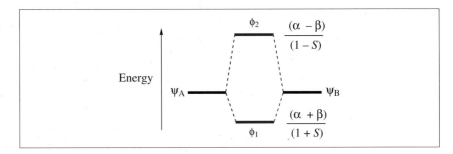

The single covalent bond, consisting of two electrons used in the bonding between two atoms, is sometimes referred to as a **two-centre two-electron (2c2e) bond**, the nomenclature being of more general use when considering larger systems.

The molecular orbitals ϕ_1 and ϕ_2 may be used to describe the electronic configurations of the helium molecule-ion, He_2^+, and the dihelium molecule, He_2. The former contains three electrons, so that two electrons pair up in the bonding orbital, leaving one unpaired electron to occupy the anti-bonding orbital. The electronic configuration of He_2^+ is $\phi_1^2\phi_2^1$. The single anti-bonding electron offsets some of the bonding effect of the pair of electrons in the bonding orbital to give a bond with a strength about equal to that of the bond in H_2^+. The extent of bonding in any system can be formalized in terms of the **bond order**. The bond order is given by the number of bonding electron pairs in excess of any anti-bonding electron pairs, or in terms of numbers of electrons as:

$$\text{Bond order} = \left(\frac{\text{No. of bonding electrons} - \text{No. of anti-bonding electrons}}{2} \right)$$

(3.38)

The electronic configuration of dihelium would be $\phi_1^2\phi_2^2$, which would result in zero bonding. Table 3.1 summarizes the electronic configurations, bond energies and bond lengths of the species H_2^+, H_2, He_2^+ and He_2.

Table 3.1 The electronic configurations and bond data for diatomic species

Molecule	Electronic configuration	Bond order	Bond dissociation energy/kJ mol^{-1}	Equilibrium bond length/pm
H_2^+	ϕ_1^1	½	264	106
H_2	ϕ_1^2	1	436	74
He_2^+	$\phi_1^2\phi_2^1$	½	297	108
He_2	$\phi_1^2\phi_2^2$	0	Molecule not formed	

The variations in the bond dissociation energies of H_2^+, H_2 and He_2^+ and their equilibrium bond lengths are consistent with the expectations from molecular orbital theory. As the bond order increases it would be expected that the bonds formed would be stronger and shorter.

3.3 Valence Bond Theory

Valence bond theory, largely developed by Pauling, is not used to a large extent in this book because the alternative molecular orbital theory, largely developed by Mulliken, gives more satisfying explanations of the bonding of molecules and in addition rationalizes their electronically excited states in a way that valence bond theory cannot easily do. Nevertheless, valence bond theory is still used in a general way to describe the bonding in many molecules, and some indication here is necessary of why it is used and its relevance to the description of the bonding in any system. Both methods aim to describe the ways in which electrons are distributed in molecules.

The molecular orbital theory of the dihydrogen molecule is dealt with in detail above, and describes how the two electrons occupy a bonding molecular orbital so that they are equally shared between the two nuclei. This state of affairs can be written symbolically in the form:

$$\phi = (\psi_a + \psi_b)(1) \times (\psi_a + \psi_b)(2) \qquad (3.39)$$

meaning that both electrons, 1 and 2, occupy the bonding molecular orbital constructed from the 1s atomic orbitals of the two hydrogen atoms, a and b. Multiplying out the two terms of the equation together produces the four-term equation:

$$\phi = \psi_a(1)\psi_a(2) + \psi_b(1)\psi_b(2) + \psi_a(1)\psi_b(2) + \psi_a(2)\psi_b(1) \quad (3.40)$$

The first two terms indicate that the two electrons occupy the 1s orbital of one or the other hydrogen atoms, *i.e.* they are the *ionic* structures, H^-H^+ and H^+H^-. The third and fourth terms indicate that each hydrogen atom is associated with one of the electrons, and are consistent with the definition of covalency. All four terms together indicate the equal sharing of the two electrons between the two nuclei, but very much overemphasize the ionic aspect of the bonding. In practice this overemphasis is dealt with by assigning coefficients to the four terms and evaluating them by the process of minimizing the total energy of the system. When minimum energy is achieved, the values of the coefficients are such that the ionic terms contribute only about 1% to the overall bonding.

Valence bond theory begins by assigning wave functions to the various permutations of electrons and nuclei to give what are known as the canonical forms of the molecule, *i.e.* for the hydrogen molecule these would be $H_a(1)H_b(2)$, $H_a(2)H_b(1)$, $H_a(1)(2)$ and $H_b(1)(2)$, terms similar to those derived from molecular orbital theory. The first two terms represent canonical forms in which each hydrogen atom has an associated electron, 1 or 2. The latter two terms are ionic in which both electrons are associated with either atom A or atom B. These canonical forms engage in **resonance interaction** (Pauling's terminology), *i.e.* they contribute to the total description of the molecular bonding such that the **resonance energy** stabilizes the system with the formation of a **resonance hybrid**. In practice, the ionic terms are given a lesser participation to arrive at a reasonable description of the bonding.

The conclusion from this short discussion is that both valence bond and molecular orbital theories can describe the bonding of a system and in the limit they both arrive at the same answer. In practice, molecular orbital theory is more often used and is much more amenable to giving solutions to more complex systems than does valence bond theory.

3.4   Energetics of the Bonding in H_2^+ and H_2

It is helpful in the understanding of covalent bond formation to consider the energies due to the operation of attractive and repulsive forces in H_2^+ and H_2, and to estimate the magnitude of the interelectronic repulsion energy in the dihydrogen molecule. Figure 3.7 shows plots of potential energy against internuclear distance for H_2^+ and H_2. The curves shown are **Morse functions** which, for diatomic species, have the form:

$$V = D_e\{1 - \exp[-(\mu/2D_e)^{1/2}\omega(r - r_{eq})]\}^2 \qquad (3.41)$$

where V represents potential energy, D_e the electronic dissociation energy, μ is the reduced mass of the system, ω is the fundamental vibration frequency of the molecule, r is the internuclear distance, and r_{eq} is the equilibrium internuclear distance otherwise known as the bond length. The quantity D_e is related to the dissociation energy, D, of the molecule by the relation:

$$D = D_e - \tfrac{1}{2}h\omega \qquad (3.42)$$

the $\tfrac{1}{2}h\omega$ term representing the **zero-point vibrational energy** of the molecule.

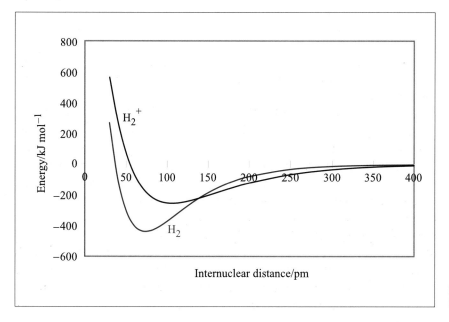

Figure 3.7 Morse curves for H_2^+ (*black line*) and H_2 (*red line*)

Under standard conditions, both molecules exist in their lowest vibrational energy levels. These are known as their **zero-point vibrational states**, in which the value of the vibrational quantum number is zero. The fact that molecules in their zero-point vibrational states possess vibrational energy is a consequence of the Uncertainty Principle; this would be violated if the internuclear distance was unchanging. The dissociation limits for both species are identical: the complete separation of the two atoms, which is taken as an arbitrary zero of energy. The difference between the zero of energy and the zero-point vibrational energy in both cases represents the bond dissociation energies, respectively, of H_2^+ and H_2.

To obtain an accurate assessment of the interelectronic repulsion energy of the H_2 molecule it is essential to carry out calculations in which the hydrogen nuclei are a constant distance apart. The following calculations are for an internuclear distance of 74 pm for both molecules, which is the equilibrium internuclear distance in the dihydrogen molecule.

3.4.1 The H_2^+ Molecule-ion

There are only two forces operating in H_2^+: the **attractive force** between the nuclei and the single electron, and the **repulsive force** between the two nuclei. The interproton repulsion energy may be calculated from Coulomb's law:

$$E(p-p) = \frac{N_A e^2}{4\pi\varepsilon_0 r} \tag{3.43}$$

where e is the charge on the proton, r is the interproton distance, ε_0 is the permittivity of a vacuum, and N_A the Avogadro constant. For two protons separated by 74 pm the force of repulsion between them causes an increase in energy of 1878 kJ mol^{-1} compared to the infinite separation of H^+ and H (the arbitrary zero of energy). From the Morse curve for H_2^+ in Figure 3.7, it may be estimated that if H^+ and H are brought from infinite separation to an interproton distance of 74 pm there is a stabilization of 180 kJ mol^{-1}. This represents the resultant energy of the system with both forces operating. It means that the attractive force operating between the electron and the two protons produces a stabilization which is in excess of 1878 kJ mol^{-1} by 180 kJ mol^{-1}, so that the quantity known as **the electronic binding energy** is calculated to be 1878 + 180 = 2058 kJ mol^{-1}. The interrelationship of these energies is shown in the diagram of Figure 3.8. Notice that the actual dissociation energy is a relatively small quantity compared to the energies representing the effects of the attractive and repulsive forces operating in the H_2^+ system.

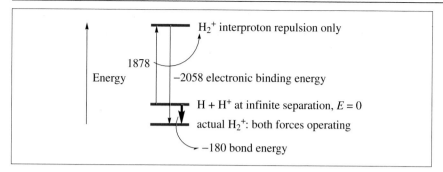

Figure 3.8 Energetics of formation of the H_2^+ molecule-ion

Worked Problem

Q What is the repusion energy between two protons which are 106 pm apart, as they are in the ground state of the hydrogen molecule-ion?

A Using equation 3.43 with $r = 106$ pm gives 1311 kJ mol^{-1} for the repulsion energy.

3.4.2 The Dihydrogen Molecule, H_2

In the dihydrogen molecule there are three forces operating: (1) interproton repulsion, (2) proton–electron attraction and (3) interelectronic repulsion. The force of interproton repulsion produces a destabilization of the dihydrogen system equal to that of the H_2^+ molecule-ion, since the interproton distance is again taken to be 74 pm. The resultant stabilization of all three forces is equal to the bond dissociation energy of H_2, which is 436 kJ mol^{-1}. For a comparison with H_2^+ the electronic binding energy may be calculated as: $1878 + 436 = 2314$ kJ mol^{-1}. It is 12% greater than that in H_2^+, indicating that *two bonding electrons are only marginally better than one at binding the two nuclei together*. The reason for this is that, with two electrons present, there is a substantial destabilization of the system as a result of the interelectronic repulsion.

The magnitude of this may be calculated, as shown in Figure 3.9, by assuming that the electronic binding energy *per electron* is as calculated for the H_2^+ system (2058 kJ mol^{-1}). For two electrons the stabilization from the electronic binding energy is $2 \times 2058 = 4116$ kJ mol^{-1}, and that amount is offset by the interelectronic repulsion energy. If the two-electron binding energy is subtracted from the sum of the two one-electron binding energies, a value of $4116 - 2314 = 1802$ kJ mol^{-1} is given for the interelectronic repulsion energy in H_2. Notice that all three energy quantities in H_2 are large by comparison with the resultant bond dissociation energy.

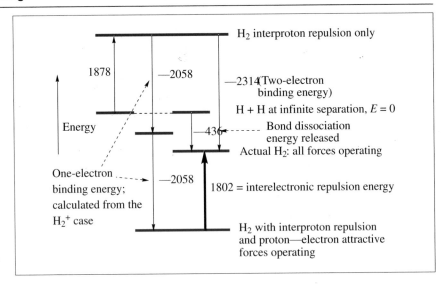

Figure 3.9 Energetics of formation of the H_2 molecule

It is of interest to calculate the magnitude of the interelectronic repulsion energy in the hydride ion, H^-, which possesses an ionic radius of 208 pm. The *electron attachment energy* of the proton is -1312 kJ mol^{-1} (numerically equal to the ionization energy of the hydrogen atom, but of opposite sign) and represents the energy released when an electron enters the 1s orbital of the hydrogen atom. The electron attachment energy of the hydrogen atom is considerably smaller than that of the proton and is -71 kJ mol^{-1}. Since the electron enters the same 1s orbital, the difference between the two electron attachment energies gives an estimate of the interelectronic repulsion energy of $-71 + 1312 = 1241$ kJ mol^{-1}. This is appreciably smaller than the value calculated for the two electrons occupying the bonding orbital of the dihydrogen molecule. The increased size of the hydride ion is one reason for this, the other being that in H_2 there are two attracting protons which draw the electrons closer to each other in spite of their like charges.

3.5   Some Experimental Observations; Photoelectron Spectroscopy

In the proper development of a scientific theory it is to be expected that theory and experimental observations should be consistent with one another, preferably the theory being refined by further observations. It is essential to obtain experimental observations by which ideas such as molecular orbital theory may be tested and possibly refined. This section describes a method of studying molecules using electromagnetic radiation.

3.5.1 Frequency, Wavelength and Energy of Electromagnetic Radiation

All electromagnetic radiation is a form of energy which is particulate in nature. The terms wavelength and frequency do not imply that radiation is to be regarded as a wave motion in a real physical sense. They refer to the form of the mathematical functions which are used to describe the behaviour of radiation. The fundamental nature of electromagnetic radiation is embodied in quantum theory, which explains all the properties of radiation in terms of **quanta** or **photons**: packets of energy. A cavity, *e.g.* an oven, emits a broad spectrum of radiation which is independent of the material from which the cavity is constructed, but is entirely dependent upon the temperature of the cavity. Planck (1918 Nobel prize for physics) was able to explain the frequency distribution of broad-spectrum cavity radiation only by postulating that the radiation consisted of quanta with energies given by the equation:

$$E = h\nu \tag{3.44}$$

where h is Planck's constant ($3.6260755 \times 10^{-34}$ J s), the equation being known as the Planck equation.

The relationship between frequency and wavelength is given by the equation:

$$\nu = c/\lambda \tag{3.45}$$

where c and λ represent the speed of light ($c = 299792458$ m s^{-1}) and its wavelength, respectively, in a vacuum. Quoted frequencies are sometimes either numbers which are too small or too large when expressed in hertz, and it is common for them to be expressed as **wavenumbers**, which are the frequencies divided by the speed of light. This is expressed by the equation:

$$\tilde{\nu} = \frac{\nu}{c} \tag{3.46}$$

Wavenumbers, represented by the symbol $\tilde{\nu}$ (nu-bar), have the units of reciprocal length and are usually quoted in cm^{-1}, although the strictly S.I. unit is reciprocal metres, m^{-1}. For example, 3650 cm^{-1} = 365000 m^{-1}. A wavenumber represents the number of full waves that occur in the unit of length. For example, the wavenumber for the stretching vibration of a C–H bond in a hydrocarbon is around 3650 cm^{-1} and means that there are 3650 wavelengths in one centimetre.

When dealing practically with equation 3.44 it is usual to work in molar quantities so that the energy has units of J mol^{-1}, and to achieve this it is necessary to use the equation:

$$E = N_A h\nu \qquad (3.47)$$

where N_A is the Avogadro constant (3.0221367×10^{23} mol^{-1}).

Worked Problem

Q What is the quantum energy of visible light of wavelength 500 nm?

A Equations 3.45 and 3.46 may be combined to give an equation for the molar quantum energy of radiation with a particular wavelength:

$$E = N_A hc/\lambda \qquad (3.48)$$

Using equation 3.47 with $\lambda = 500 \times 10^{-9}$ m gives the molar quantum energy of 500 nm radiation as 239.2 kJ mol^{-1}. A mole of quanta is known as an **Einstein**

3.5.2 Photoelectron Spectroscopy

A direct method of obtaining experimental measurement of the energies of electrons in molecules is known as **photoelectron spectroscopy (PES)**. The basis of the method is to bombard an atomic or molecular species with radiation of sufficient energy to cause its ionization. If the quantum energy of the radiation is high enough, ionizations may be caused from one or other of the permitted levels within the bombarded atom or molecule. In addition to causing ionization of the target species, the radiation (in the case of molecules) may cause changes in the vibrational, ΔE_{vib}, and rotational, ΔE_{rot}, energies of the resulting positive ion. This may be written as:

$$M(g) + h\nu \rightarrow M^+ + e^- \text{ (photoelectron)} \qquad (3.49)$$

the energy balance being:

$$h\nu = I_M + \Delta E_{vib} + \Delta E_{rot} + \text{kinetic energy of the electron} \qquad (3.50)$$

The kinetic energies (KEs) of the photoelectrons are measured by the use of a modification of a conventional β-ray spectrometer as used in the study of β-particle (electron) emissions from radioactive nuclei. The photoelectron spectrum of a molecule is presented as a plot of the count rate (the intensity of the photoelectrons detected) against the electron energy. The PES of the dihydrogen molecule is shown in Figure 3.10.

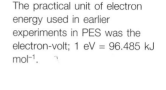

The practical unit of electron energy used in earlier experiments in PES was the electron-volt; 1 eV = 96.485 kJ mol⁻¹.

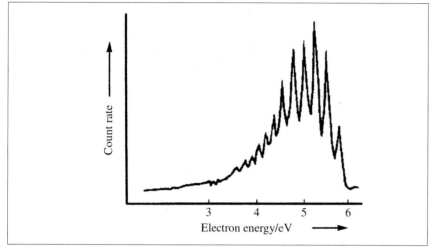

Figure 3.10 The photoelectron spectrum of the H_2 molecule

The energy of the quantum used to cause the ionization is 21.22 eV or 2047 kJ mol⁻¹, so the ionization energy of the dihydrogen molecule is 21.22 eV minus the energy of the photoelectron with the energy of 5.8 eV, corresponding to the peak on the far right-hand side of the spectrum, which gives 15.42 eV as the first ionization energy of H_2. The additional peaks in the photoelectron spectrum represent ionizations of the molecule, with additional energy being used to excite the product H_2^+ molecule-ion vibrationally and rotationally. The rotational 'fine structure', which arises from variations in the rotational energy of the ion and its neutral molecule (ΔE_{rot}), is not observed at the resolution at which the spectrum was measured, but contributes to the width of the vibrational bands which are observed. Vibrational excitations up to the eleventh level can be discerned from the spectrum shown.

The second peak from the right in the dihydrogen PES corresponds to the energy required for its ionization plus that required to give H_2^+ one quantum of vibrational excitation. The difference in energy between the first two peaks is a measure of the magnitude of one quantum of vibrational excitation energy of H_2^+. The difference amounts to 0.3 eV or 29 kJ mol⁻¹. It is useful to compare such a value with the energy of one quantum of vibrational excitation of the dihydrogen molecule, which is 49.8 kJ mol⁻¹. The frequency (and in consequence the energy) of the vibration of two atoms bonded together is related to the bond strength, so it may be concluded that the removal of an electron from a dihydrogen molecule causes the remaining bond (in H_2^+) to be considerably weaker than that in the parent molecule. This is a confirmation of the bonding nature of the electron removed in the ionization process. Diagnosis of the bonding, non-bonding or anti-bonding nature of electrons in molecules may in some cases be made from a study of the effects of their ionization upon the vibrational frequency of the resulting positive ion. The removal of a non-bonding electron would have very little effect upon the bond strength and the vibrational frequencies of the neutral molecule, and its unipositive ion would be expected to be similar in bond strength and vibrational frequency. The vibrational frequency of the ion would be greater than that of the neutral molecule if the electron removed originally occupied an anti-bonding orbital.

3.5.3 Photodissociation of the Dihydrogen Molecule

Experimental confirmation of the energy of the anti-bonding level in the dihydrogen molecule comes from the observation of its absorption spectrum in the far-ultraviolet region. Dihydrogen absorbs radiation of a wavelength of 109 nm, which is equivalent to a molar quantum energy of 1097 kJ mol^{-1}.

In terms of an **electronic transition** the process of absorbing the quantum is:

$$\phi_1^2 (\text{ground state configuration}) \xrightarrow{\ h\nu\ } \phi_1^1 \phi_2^1 (\text{excited state configuration})$$

$$(3.51)$$

which would lead to the dissociation of the molecule into two hydrogen atoms. The effect of the bonding electron in ϕ_1 is out-balanced by the anti-bonding electron in ϕ_2 and the molecule undergoes photodissociation to give two hydrogen atoms.

Summary of Key Points

1. The differences between covalent, dative or coordinate, and ionic bonding were explained.

2. Molecular orbital theory was introduced by the use of linear combinations of atomic orbitals (LCAO).

3. Group theory was used to define the orbitals of the dihydrogen molecule. This led to the molecular orbital diagram for H_2 and the relative energies of atomic, bonding and anti-bonding orbitals.

4. Filling the bonding molecular orbital led to the concept of the single covalent bond.

5. Valence bond theory was introduced and compared to molecular orbital theory.

6. The energetics of the dihydrogen molecule-ion, H_2^+, and the dihydrogen molecule were outlined to give a calculation of the inter-electronic repulsion energy in the latter.

7. The photoelectron spectrum of the dihydrogen molecule was described and interpreted in terms of molecular orbital theory.

Problems

3.1. What atomic orbitals of the hydrogen atom are used to construct the molecular orbitals of the dihydrogen molecule?

3.2. Draw the overlap diagram for the bonding orbital in the dihydrogen molecule.

3.3. Draw the overlap diagram for the anti-bonding orbital in the dihydrogen molecule.

3.4. What are the symmetry symbols for the bonding and anti-bonding orbitals of H_2?

3.5. What property of a bonding molecular orbital is mainly responsible for its bonding properties?

3.6. Write down the electronic configuration of the He_2^+ ion and state the bond order.

3.7. Describe the normalization process applied to a molecular orbital.

3.8. Describe the difference between covalent and ionic bonds.

3.9. Describe what is meant by an overlap integral.

3.10. The bond dissociation energies of H_2 and H_2^+ are 436 and 264 kJ mol^{-1}, respectively. Correlate these data with the bond order of the species, and comment on the equilibrium bond lengths of 74 and 106 pm, respectively.

3.11. Describe the photoelectron spectrum of the H_2 molecule and indicate what information can be gained about the H_2^+ ion.

Further Reading

D. M. P. Mingos, *Essentials of Inorganic Chemistry 1*, Oxford University Press, Oxford, 1995. This is an A–Z account of the major concepts used in modern inorganic chemistry.

D. W. Turner, C. Baker, A. D. Baker and C. R. Brundle, *Molecular Photoelectron Spectroscopy*, Wiley, New York, 1969. A book full of useful spectra and their interpretation. Out of print, but available in chemistry libraries.

4

Covalent Bonding II: Diatomic Molecules; Bond Lengths and Strengths

---

This chapter consists of the application of the symmetry concepts of Chapter 2 and those of molecular orbital theory to the construction of molecular orbitals for a range of diatomic molecules, building upon the principles outlined in Chapter 3. These are: (1) the homonuclear diatomic molecules, E_2, where E is an element of the second period of the periodic classification (Li–Ne), and (2) some heteronuclear diatomic molecules, in which differences in electronegativity coefficients between the combining atoms are important: nitrogen monoxide, carbon monoxide and hydrogen fluoride. Emphasis is placed on the length and strength of bonds in relation to the electronic configurations of the molecules.

Aims

By the end of this chapter you should understand:

- The construction of the molecular orbitals of the homonuclear diatomic molecules of the second period using group theory
- Their bond lengths and strengths and the relationship of these quantities with the electronic configurations of the molecules
- What bond order is
- The effects of electronegativity differences on the bonding in heteronuclear diatomic molecules
- The photoelectron spectra of selected diatomic molecules

4.1 Homonuclear Diatomic Molecules of the Second Short Period Elements, E_2

This section consists of the formal molecular orbital treatment of homonuclear diatomic molecules of the second short period. The molecular axis of an A_2 molecule is arranged to be coincident with the z axis by convention. It is necessary to look at the 2s and 2p orbitals separately and classify them with respect to the $D_{\infty h}$ point group to which the molecules belong.

4.1.1 Classification of the Orbitals of E$_2$ Molecules

Classification of the 2s Orbitals of E$_2$ Molecules

The classification of the two 2s orbitals of an E_2 molecule is very similar to that of the two 1s orbitals of dihydrogen and is not repeated here. The two combinations of the 2s orbitals are:

$$\phi(2\sigma_g^+) = \psi(2s)_A + \psi(2s)_B \tag{4.1}$$

and

$$\phi(2\sigma_u^+) = \psi(2s)_A - \psi(2s)_B \tag{4.2}$$

the A and B subscripts referring to the two atoms contributing to the molecule. To simplify the equations it is assumed that all the wave functions are normalized, although the normalization factors are omitted. The equations are used to indicate the atomic orbitals which contribute to the molecular orbitals.

There is a bonding combination (equation 4.1) which transforms within the $D_{\infty h}$ point group as a σ_g^+ irreducible representation, the prefix 2 being assigned because of the bonding combination of the 1s orbitals having the same symmetry (and termed $1\sigma_g^+$). Likewise the anti-bonding combination (equation 4.2) is termed $2\sigma_u^+$ because there is an anti-bonding combination of the 1s orbitals at lower energy, $1\sigma_u^+$.

The use of numerical prefixes to distinguish between orbitals of the same symmetry was suggested by Mulliken and is the generally accepted method. The prefixes may be omitted when it is clear that no ambiguity is present.

Classification of the 2p Orbitals of E$_2$ Molecules

The character of the reducible representation of the 2p orbitals of E_2 molecules may be obtained by writing down, under each symmetry element of the $D_{\infty h}$ group, the number of such orbitals which are unchanged by each symmetry operation. This produces the representation:

	E	C_∞^ϕ	σ_v	i	S_∞^ϕ	C_2
$6 \times 2p$	6	$2 + 4\cos\phi$	2	0	0	0

This result requires some explanation, particularly concerning the character of the 2p orbitals with respect to the C_∞ operation. The two $2p_z$ orbitals, lying along the C_∞ axis with their positive lobes overlapping, are unaffected by the associated operation and account for the 2 in the character column. The term $4\cos\phi$ arises because of the two $2p_x$ and two $2p_y$ orbitals which are perpendicular to the C_∞ axis. Although a rotation through ϕ degrees around that axis does not move any of the orbitals to another centre, it does alter their disposition with regard to the xz and yz planes.

If the angle ϕ was chosen to be $180°$, for instance, it would have the effect of inverting the $2p_x$ and $2p_y$ orbitals, and it would be necessary to place -1 in the above representation for each orbital as their characters (note that $\cos 180° = -1$). To take into account all possible values of ϕ it is essential to express the character of each orbital as the cosine of the angle of rotation, ϕ. Effectively this implies that the character of each orbital is represented by the resolution of the orbital on to the plane it occupied before the symmetry operation was carried out. This ensures that for a rotation through $180°$ the character of a $2p_x$ or $2p_y$ orbital will be -1; in effect such an orbital, whilst not moving from its original position, changes the signs of its ψ values.

The character with respect to reflexion in one of the infinite number of vertical planes requires some explanation. It is best to choose a particular vertical plane such as that represented by the xz plane. Reflexion in any of the vertical planes has no effect upon the two $2p_z$ orbitals, which gives 2 as their character. Reflexion of the two $2p_x$ orbitals in the xz plane does not change them in any way: their character is 2. The reflexion of the two $2p_y$ orbitals in the xz plane causes their ψ values to change sign, and because they are otherwise unaffected, their character is -2. The resultant character of the six $2p$ orbitals, with respect to the operation, σ_v, is given by $2 + 2 - 2 = 2$.

The reducible representation of the six $2p$ orbitals may be seen, by inspection of the $D_{\infty h}$ character table and carrying out the following exercise, to be equivalent to the sum of the irreducible representations:

$$6 \times 2p = \sigma_g^+ + \sigma_u^+ + \pi_g + \pi_u \tag{4.3}$$

Worked Problem

Q Demonstrate the truth of equation 4.3 by summing the characters of the appropriate representations from the $D_{\infty h}$ character table (Appendix 1).

A The required sum is:

	E	C_∞^ϕ	σ_v	i	S_∞^ϕ	C_2
σ_g^+	1	1	1	1	1	1
σ_u^+	1	1	1	−1	−1	−1
π_g	2	$2\cos\phi$	0	2	$-2\cos\phi$	0
π_u	2	$2\cos\phi$	0	−2	$2\cos\phi$	0
Sum	6	$2+4\cos\phi$	2	0	0	0

and is seen to be equal to the deduced reducible representation of the six 2p orbitals given above.

The convention used throughout this text to express the form of a molecular orbital is to use a *plus sign* in the equations to indicate a *bonding* combination and a *minus sign* to indicate an *anti-bonding* combination.

It is important to realize which orbital combinations are represented by the above irreducible representations. This is best achieved by looking at the diagrams in Figure 4.1 for the overlaps represented by the equations:

$$\phi(3\sigma_g^+) = \psi(2p_z)_A + \psi(2p_z)_B \qquad (4.4)$$

$$\phi(3\sigma_u^+) = \psi(2p_z)_A - \psi(2p_z)_B \qquad (4.5)$$

$$\phi(1\pi_u) = \psi(2p_x)_A + \psi(2p_x)_B; \qquad \psi(2p_y)_A + \psi(2y_z)_B \qquad (4.6)$$

$$\phi(1\pi_g) = \psi(2p_x)_A - \psi(2p_x)_B; \qquad \psi(2p_y)_A - \psi(2y_z)_B \qquad (4.7)$$

The molecular orbital $\phi(3\sigma_g^+)$ is bonding and is the *third* highest energy σ_g^+ orbital; hence the prefix '3'. The $\phi(3\sigma_u^+)$ orbital is the anti-bonding combination of the two $2p_z$ orbitals, and the third highest energy σ_u^+ orbital. The π_u and π_g orbitals are both *doubly degenerate*, the π_u combination being bonding, the π_g combination being anti-bonding. They both are prefixed by the figure '1' since they are the lowest energy orbitals of their type.

In the discussion of the bonding of the E_2 molecules in the next section, the orbitals are referred to by their symmetry symbols with the appropriate numerical prefixes.

4.1.2 The Molecular Orbital Diagram for E_2 Molecules

Figure 4.2 is a diagram of the relative energies of the molecular orbitals of the E_2 molecules, together with those of the atomic orbitals from which they were constructed. The 'sideways' overlap involved in the production of π orbitals is not as effective as the 'end-on' overlap which characterizes the production of σ orbitals. For a given interatomic dis-

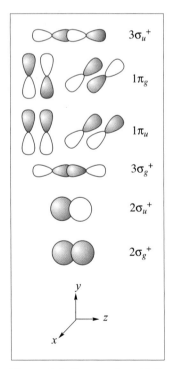

Figure 4.1 The σ and π orbital overlaps between 2p atomic orbitals in the formation of some diatomic molecules

tance the overlap integral for σ-type overlap is generally higher than that for π-type overlap between two orbitals. The consequence of this is that the bonding stabilization and the anti-bonding destabilization associated with π orbitals are significantly less than those associated with σ orbitals. This accounts for the differences in energy, shown in Figure 4.2(a), between the σ and π orbitals which originate from the 2p atomic orbitals of the E atoms. The order of energies of the orbitals of E_2 molecules according to Figure 4.2(a) is dependent upon the assumption that the energy difference between the 2p and 2s atomic orbitals is sufficient to prevent significant interaction between the molecular orbitals. In Figure 4.2(b), interactions are allowed between molecular orbitals of the same symmetry, which produces a different *aufbau* filling order. This effect is more important when the 2p–2s energy gap is relatively small.

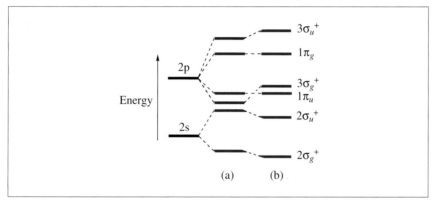

Figure 4.2 The molecular orbital diagrams for homonuclear diatomic molecules of the second short period, Li_2 to Ne_2. Diagram (a) is appropriate for O_2, F_2 and Ne_2, diagram (b) for the molecules Li_2 to N_2

Box 4.1 Equations for Overlap Integrals

Mulliken derived equations for the variation of overlap integrals with distance between the participating nuclei. For homonuclear diatomic molecules the equations for the overlap of two 2p orbitals in σ and π modes are respectively:

$$S(2p_\sigma - 2p_\sigma) = e^{-p}[-1 - p - {}^1\!/_5 p^2 + {}^2\!/_{15} p^3 + {}^1\!/_{15} p^4] \qquad (4.8)$$

$$S(2p_\pi - 2p_\pi) = e^{-p}[1 + p + {}^2\!/_5 p^2 + {}^1\!/_{15} p^3] \qquad (4.9)$$

In these equations the *p* term is the product of the effective nuclear charge, derived by the application of Slater's rules (see Box 4.2), and the distance apart of the two nuclei, *R*, in atomic units, *i.e.* $Z_{eff} R/2a_0$. The two functions are plotted against *p* in Figure 4.3, which shows that π overlap becomes more and more efficient as the internuclear distance is reduced, but the σ overlap moves through a maximum and eventually becomes negative at small distances.

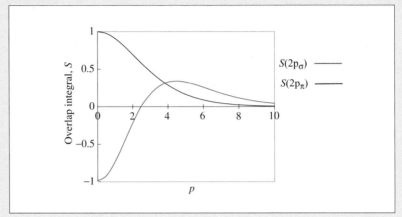

Figure 4.3 Plots of the overlap integrals against the parameter p for the σ and π overlaps between pairs of 2p atomic orbitals

This is where the positive and negative portions of the σ orbitals start to overlap. Ranged against the maximization of the overlap integrals is the internuclear repulsion, and the equilibrium position is attained when there is a balance between the forces of attraction and repulsion.

Box 4.2 Slater's Rules

The effective atomic number is considered to be the difference between the actual atomic number, Z, and a screening constant, S, which is estimated by the use of Slater's rules. These represent an approximate method of calculating the value of S, leading to an estimate of the effective atomic number, Z_{eff}, given by $Z - S$. The value of S is obtained from the following rules.

1. The atomic orbitals are divided into groups: (1s); (2s, 2p); (3s, 3p); (3d); (4s, 4p); (4d); (4f); (5s, 5p); (5d); (5f); and so on.

2. For an electron (principal quantum number n) in a group of s and/or p electrons, the value of S is given by the sum of the following contributions:
(i) Zero from electrons in groups further away from the nucleus than the one considered.
(ii) 0.35 from each other electron in the same group, unless the group considered is the 1s when an amount 0.30 is used.

(iii) 0.85 from each electron with principal quantum number of $n - 1$.

(iv) 1.00 from all other inner electrons.

3. For an electron in a d or f group the parts (i) and (ii) apply, as in rule 2, but parts (iii) and (iv) are replaced with the rule that all the inner electrons contribute 1.00 to S.

Worked Problems

Q Calculate the screening constant for an electron in the 2p orbital of the oxygen atom.

A The electronic configuration of the oxygen atom is $1s^2 2s^2 2p^4$. To operate Slater's rules these electrons are divided into two groups, $1s^2$ and $2s^2 2p^4$. For an electron in a 2p orbital there are five other electrons in the group which each contribute 0.35 to the screening constant, *i.e.* $5 \times 0.35 = 1.75$. The two 'inner' electrons both contribute 0.85 to the screening constant, *i.e.* $2 \times 0.85 = 1.7$. The value of S for the 2p electron is $1.75 + 1.7 = 3.45$, so the effective nuclear charge is $8 - 3.45 = 4.55$.

Q Why are σ-bonds *generally* stronger than π-bonds?

A The p values for the molecules B_2, C_2, N_2, O_2 and F_2 are calculated from the internuclear distances and the Slater values for the effective nuclear charges of the respective atoms:

Molecule	$Z_{eff} = Z - S$	$p = Z_{eff} R / 2a_0$
B_2	$5 - 2.4 = 2.6$	3.98
C_2	$6 - 2.75 = 3.25$	3.88
N_2	$7 - 3.1 = 3.9$	4.11
O_2	$8 - 3.45 = 4.55$	5.29
F_2	$9 - 3.8 = 5.2$	7.2

As may be seen from Figure 4.3, the p values for most π-bonded molecules are mainly in the region where the overlap integral for the σ overlap is significantly larger than that for the π overlap.

If two molecular orbitals have identical symmetry, such as the $2\sigma_g^+$ and $3\sigma_g^+$ orbitals, they may interact by the formation of linear combinations. The resulting combinations still have the same symmetry (and retain the nomenclature), but the lower orbital is stabilized at the expense of the upper one. Such interaction is also possible for the $2\sigma_u^+$ and $3\sigma_u^+$ orbitals. The extent of such interaction is determined by the energy gap between the two contributors. If the energy gap between the 2p and 2s atomic orbitals is small enough, the interaction between the $2\sigma_g^+$ and $3\sigma_g^+$ molecular orbitals may be so extensive as to cause the upper orbital $3\sigma_g^+$ to have an energy which is greater than that of the $1\pi_u$ set. Such an effect is shown in Figure 4.2(b).

The magnitude of the 2p–2s energy gap varies along the elements of the second period (Li–Ne), as is shown in Figure 4.4. The energy gaps in the elements lithium to nitrogen are sufficiently small to make significant $2\sigma_g^+$ and $3\sigma_g^+$ interaction possible, such that Figure 4.2(b) is relevant in determining the electronic configurations of the molecules Li_2, Be_2, B_2, C_2 and N_2. Figure 4.2(a) is used to determine the electronic configurations of the molecules O_2, F_2 and Ne_2, since the 2p–2s energy gaps in O, F and Ne are sufficiently large to prevent significant molecular orbital interaction. The *aufbau* principle is followed in building up the electronic configurations of the diatomic molecules, with the relevant order of filling given by the appropriate part of Figure 4.2.

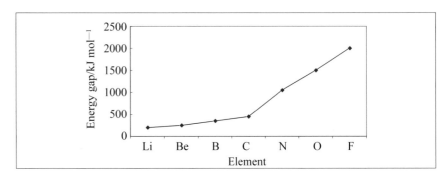

Figure 4.4 Variation of the 2p–2s energy gap along the second short period

The interaction of molecular orbitals of the same symmetry has important consequences for all systems where it occurs, and more examples will be referred to in later chapters. It is possible to carry out the mixing of the original atomic orbitals (known as **hybridization**) before the molecular orbitals are formed. Some examples of this approach are included in later chapters. Both approaches give the same eventual result for the contributions of atomic orbitals to the molecular ones.

4.2 Electronic Configurations of Homonuclear Diatomic Molecules

The electronic configurations of the homonuclear diatomic molecules of the elements of the second period, and some of their ions, are given in Table 4.1.

Table 4.1 The electronic configurations of some homonuclear diatomic molecules and ions

Molecule/ion	$2\sigma_g^+$	$2\sigma_u^+$	$3\sigma_g^+$	$1\pi_u$	$1\pi_g$	$3\sigma_u^+$
Li_2	2					
Be_2	2	2				
B_2	2	2		2		
C_2	2	2		4		
N_2	2	2	2	4		
N_2^+	2	2	1	4		
N_2^-	2	2	2	4	1	
O_2	2	2	2	4	2	
O_2^+	2	2	2	4	1	
O_2^-	2	2	2	4	3	
O_2^{2-}	2	2	2	4	4	
F_2	2	2	2	4	4	
Ne_2	2	2	2	4	4	2

The four electrons occupying the $1\sigma_g^+$ and $1\sigma_u^+$ orbitals are not indicated. The interactions of the 1s orbitals of the atoms under discussion are minimal and, although they may be regarded as occupying the molecular orbitals previously indicated, they are virtually non-bonding. In any case, the slight bonding character of the two electrons occupying $1\sigma_g^+$ is cancelled by the slight anti-bonding nature of the two electrons in $1\sigma_u^+$. In some texts the four electrons are indicated symbolically by KK (a reference to the 'K shells' of the two atoms), now recognized as their 1s atomic orbitals.

4.2.1 Dilithium, Li_2

This molecule exists in the gas phase and has a bond dissociation energy of 107 kJ mol^{-1} and a bond length of 267 pm. The weak, and very long, bond is understandable in terms of the two electrons in $2\sigma_g^+$ being the only ones having bonding character. The bond order is one. The four 1s electrons have no resultant bonding effect and yet contribute considerably to the interelectronic repulsion energy. The electron attachment energy of the lithium atom is -59.8 kJ mol^{-1}, which indicates that the nuclear charge of $+3e$ is not very effective in attracting more electrons. The atom also has a very low first ionization energy (513 kJ mol^{-1}), which is another indication of the low effectiveness of the nuclear charge. Another approach is to use the concept of the nuclear charge being shielded by the various occupied atomic orbitals so as to reduce its effectiveness in attracting extra electrons. *For bonding to be achieved, the attraction between the two shielded nuclei and the two bonding electrons must outweigh the two repulsive interactions: internuclear and interelectronic.* The efficient shielding of the lithium nuclei by their $1s^2$ 'core' configurations contributes to the weakness of the bond in the Li$_2$ molecule. The standard state, *i.e.* the more thermodynamically stable form, of lithium is the metallic state, which does not contain discrete molecules and has a metallic lattice. The bonding in metals is discussed in Chapter 7.

4.2.2 Diberyllium, Be$_2$

The electronic configuration of the diberyllium molecule, Be$_2$, would be $(2\sigma_g^+)^2(2\sigma_u^+)^2$, with two bonding electrons being counterbalanced by two anti-bonding electrons, leading to a bond order of zero. The molecule does not exist with such a configuration. Elementary beryllium exists in the solid state with a metallic lattice.

4.2.3 Diboron, B$_2$

The diboron molecule, B$_2$, has a transient existence in the vapour of the element, and has a bond dissociation energy of 291 kJ mol^{-1}, the bond length being 159 pm. The bond is stronger and shorter than that in Li$_2$. The first ionization energy of the boron atom (800 kJ mol^{-1}) indicates that the nuclear charge is considerably more effective than that of the lithium atom and a somewhat stronger (and shorter) bond is to be expected for B$_2$ as compared to that in Li$_2$. The two pairs of σ electrons have a zero resultant bonding effect, leaving the stability of the bond to the two electrons which occupy the $1\pi_u$ orbitals. It is of interest that, since the $1\pi_u$ level is doubly degenerate (*i.e.* Hund's rules apply to their filling), the two orbitals are singly occupied and the two π_u electrons have parallel spins. The bonding in B$_2$ consists of two 'half-π' bonds, if the term 'bond' is understood to indicate a pair of bonding electrons. The bond order is 1.0.

Evidence which is consistent with the above description of the bonding in B$_2$ is the observation that the molecule is **paramagnetic**, the property associated with an unpaired electron (or with more than one unpaired electrons with parallel spins). In the B$_2$ case the evidence for the presence of unpaired electrons comes from the observation of its electron spin resonance (ESR) spectrum. The detailed theory of electron spin resonance spectra is not dealt with in this book. The essentials of the method depend upon there being a difference in the energy of an unpaired electron when subjected to a magnetic field. The electron spin is aligned either in the same direction as the applied field or against it. The difference in energy between the two quantized alignments corresponds to the energies of radiofrequency photons. No ESR signal is obtained from paired-up electrons, since neither of the electrons may change its spin without violating the Pauli exclusion principle.

Elementary boron exists in the solid state in several allotropic forms. Those which have been properly characterized by X-ray diffraction contain icosahedral B_{12} units, which are also the basis of many boron cluster compounds.

4.2.4 Dicarbon, C_2

This molecule exists transiently in hydrocarbon/dioxygen flames and has a bond dissociation energy of 590 kJ mol^{-1} and a bond length of 124 pm. The bond order is 2, since both the orbitals of the bonding $1\pi_u$ set are filled. The atoms are held together by two π bonds with no overall σ-bonding: a very unusual example. The carbon nucleus is more effective than that of the boron atom and, combined with there being twice as many resultant bonding electrons, serves to produce a much more stable molecule (with respect to the constituent atoms) than in the case of B_2. The average bond energy for C–C σ bonds is generally accepted to be 348 kJ mol^{-1}, and that for C=C double ($\sigma + \pi$) bonds is 612 kJ mol^{-1}. These energies are consistent with the view expressed above that π bonding is weaker than σ. The extra π-bond energy of the C=C double bond is given by 612 – 348 = 264 kJ mol^{-1}, and is about 25% smaller than that representing the strength of the C–C single σ bond.

Elementary carbon exists as the allotropes diamond, graphite and the recently characterized series of cluster molecules known as fullerenes, *e.g.* C_{60}.

4.2.5 Dinitrogen, N_2, and the Ions N_2^+ and N_2^-

In the dinitrogen molecule, N_2, two electrons occupy the $3\sigma_g^+$ orbital, and the bond order is 3.0: one σ pair plus the two π pairs. The electronic configuration is consistent with the very high bond dissociation energy of 942 kJ mol^{-1} and the short bond length of 109.7 pm. In the sequence of diatomic molecules, dinitrogen is the first that represents the standard state of the element. The molecule is chemically inert to oxidation and reduction, although it does easily form some complexes when it acts as a ligand as, for example, in $[Ru(NH_3)_5N_2]^{2+}$. The great strength of the bond in dinitrogen is associated with the presence of an excess of six bonding electrons together with the greater effectiveness of the nuclear charge compared to that of carbon.

The ionization of the molecule to give the N_2^+ ion causes the bond order to be reduced to 2.5, with consequent weakening (bond dissociation energy = 841 kJ mol^{-1}) and lengthening (bond length = 111.6 pm) of the bond as compared to that in N_2. The electron removed in the ionization comes from the $3\sigma_g^+$ orbital, which, because of the interaction with the $2\sigma_g^+$ orbital, is only moderately bonding. The effects upon the bond strength and length are, therefore, relatively small.

The N_2^- ion is produced by adding an electron to the anti-bonding $1\pi_g$ level and, as does the N_2^+ ion, it has a bond order of 2.5. Since the $1\pi_g$ level is anti-bonding, with no off-setting effects, the addition of an electron causes the bond in N_2^- to be significantly weaker (bond dissociation energy = 765 kJ mol^{-1}) and longer (bond length = 119 pm) than the one in N_2.

Worked Problem

Q The carbide ion, C_2^{2-}, is a constituent of some compounds of electropositive metals, such as in CaC_2. Write down the electronic configuration of the carbide ion and determine the bond order. The C–C distance in CaC_2 is 119 pm. Is this distance compatible with your value for the bond order?

A The carbide ion is isoelectronic with the dinitrogen molecule and has the same electronic configuration, $(2\sigma_g^+)^2(2\sigma_u^+)^2(1\pi_u)^4(3\sigma_g^+)^2$, and, in consequence, the same bond order of 3. The bond length should be small. The bond length is not as small as in the dinitrogen case, but that is probably because of the double negative charge, the repulsions contributing to a slight increase in the bond length.

4.2.6 Dioxygen, O_2, and the Ions O_2^+, O_2^- and O_2^{2-}

The standard state of oxygen is the dioxygen molecule, O_2, which has the dinitrogen configuration (except that the $1\pi_u$ level is higher in energy than the $3\sigma_g^+$ orbital) with an extra two electrons which occupy the $1\pi_g$ level. Since $1\pi_g$ is doubly degenerate, the orbitals are singly occupied by electrons with parallel spins (in accordance with Hund's rules). The molecule is paramagnetic as would be expected, and since the additional electrons occupy anti-bonding orbitals, the bond order decreases to 2.0 compared with the N_2 molecule. In consequence, the bond dissociation energy (494 kJ mol^{-1}) is considerably lower, and the bond length (121 pm) is considerably larger, than in the N_2 case.

The ionization of an electron from the highest energy level ($1\pi_g$) of the O_2 molecule produces the positive ion, O_2^+, which has a bond order of 2.5 (as does N_2^-). The bond dissociation energy of O_2^+ is 644 kJ mol^{-1} and its bond length is 112 pm. The nuclear charge of the oxygen is not as effective as that of nitrogen, and the anti-bonding electron in O_2^+ has a very significant weakening effect.

In the superoxide ion, O_2^-, there are three electrons occupying the $1\pi_g$ orbitals. The bond order is 1.5, which is consistent with its observed bond dissociation energy of 360 kJ mol^{-1} and bond length of 132 pm.

The peroxide ion, O_2^{2-}, has a filled set of $1\pi_g$ orbitals and the bond is weaker (bond dissociation energy = 149 kJ mol^{-1}) and longer (bond length = 149 pm) than that of O_2^-. The order of the bond in the O_2^{2-} ion is 1.0.

Valence bond theory is not discussed in this chapter. Molecular orbital theory is so superior in this area in that it succeeds in describing the electronic and magnetic properties and the bonding of the dioxygen molecule. The valence bond approach to O_2 would be to assume that two electron pairs were shared between the two atoms to give the structure :Ö::Ö:, which has all the electrons paired up and would not explain the paramagnetism of the molecule, nor would the changes in bond lengths and strengths on ionization be explained.

Worked Problems

Q In the compound V_3B_2 the boron atoms exist in pairs, suggesting that the B_2 unit is present with B–B bonds. The inter-boron distance is 173 pm. Is this bond distance compatible with a B_2 unit?

A If the oxidation state of the vanadium atoms in V_3B_2 is +2, then that of the boron atoms is –3. An ion with the formula B_2^{6-} would be isoelectronic with the dioxygen molecule and would have a bond order of 2. The relatively large bond length in B_2^{6-} is the result of the very high charge, which would be associated with a large contribution to the interelectronic repulsion energy. Additionally, it should be appreciated that the conclusion arrived at is dependent upon the ionic nature of the compound. It is by no means certain that this is the case and the B_2 units probably do not have as much negative charge as is assumed for the ionic case.

Q The S–S distance in the molecule S_2 is 189 pm. Determine the electronic configuration of this molecule and calculate the bond order. Would you expect the molecule to be paramagnetic?

A S_2 is isoelectronic with O_2 and therefore has a bond order of 2. The molecule should be paramagnetic since it has two unpaired electrons in the π_g anti-bonding orbitals.

Q The ion S_2^- contributes to the colour of the mineral *lapis lazuli* and synthetic materials of a similar structure called *ultramarines*.

The deep blue colour of *lapis lazuli* is due to the S_3^- radical-ion, but in the ultramarines the colour ranges from deep blue to green; the green materials containing a greater proportion of S_2^- ions.

Calculate the bond order in the ion, and decide whether the S–S distance should be greater or smaller than that in S_2.

A The ion S_2^- has a bond order of 1.5 and should have an S–S distance larger than 189 pm. In the ultramarines the S–S distance falls into the range 208–215 pm.

4.2.7 Difluorine, F_2

The standard state of fluorine is the difluorine molecule, F_2, which has an electronic configuration identical with that of the peroxide ion. The two species are isoelectronic. The bond order is 1, and the bond dissociation energy of 155 kJ mol^{-1} and bond length of 144 pm are very similar to the values for O_2^{2-}.

The species with any electrons occupying the $1\pi_g$ (anti-bonding) level (O_2^+, O_2, O_2^-, O_2^{2-} and F_2) possess considerable chemical reactivity. The bonds are relatively weak and, therefore, easily cleaved, and the species with unpaired electrons can easily form linkages to other atoms.

4.2.8 Dineon, Ne_2

The standard state of neon is the gaseous atom. The dineon molecule, Ne_2, with all its molecular orbitals filled, has an equal number of bonding and anti-bonding orbitals doubly occupied, resulting in a bond order of zero, and would not be expected to exist.

Worked Problems

Q Which elements exist in their *standard states* as homonuclear diatomic molecules?

A Only nitrogen, oxygen and the halogen (Group 17) elements. They are electronegative non-metals.

Q The molecule Ne_2 does not exist. Determine the bond order of the Ne_2^+ ion and comment upon its possible existence. Under which conditions might the ion be produced?

A   The Ne_2^+ ion would have the electronic configuration $(2\sigma_g^+)^2(2\sigma_u^+)^2(1\pi_u)^4(3\sigma_g^+)^2(1\pi_g)^4(3\sigma_u^+)^1$, leading to a bond order of 0.5. It has some stability, therefore, with respect to the separated

Ne atom and Ne^+ ion. It does exist transiently in a discharge tube in which neon gas is present. [In 1997 it was confirmed that the Xe_2^+ ion existed in the compound $Xe_2Sb_4F_{21}$ with an Xe–Xe bond length of 308.7 pm, a long bond as expected for a bond order of 0.5]

4.2.9 Bond Order, Bond Length and Bond Strength Relationships

The bond orders, bond lengths and bond dissociation energies for the diatomic species discussed in the previous section are summarized in Table 4.2.

Table 4.2 Bond orders, bond lengths and bond dissociation energies for some diatomic species

Diatomic species	Bond order	Bond dissociation energy/kJ mol^{-1}	Bond length/pm
Li_2	1	267	107
Be_2	0	–	–
B_2	1	291	159
C_2	2	590	124
N_2	3	942	109.7
N_2^+	2.5	841	111.6
N_2^-	2.5	765	119
O_2	2	494	121
O_2^+	2.5	644	112
O_2^-	1.5	360	132
O_2^{2-}	1	149	149
F_2	1	159	144
Ne_2	0	–	–

There is only a very general relationship between bond order and bond length and between bond order and bond strength as measured by the dissociation energy of a molecule. A plot of the bond lengths and bond strengths of the dioxygen species (given in Table 4.2) against bond order is shown in Figure 4.5. This shows that the relationship is almost linear, but it would seem that the O–O single bond is slightly longer than would be expected from the other values. The single bond in F_2 is also particularly long, and this is usually rationalized in terms of a large contribution from the repulsions between the three non-bonding pairs of electrons on each fluorine atom. This is similar to the explanation of the magnitude of the electron attachment energy of the fluorine atom. It may be argued that the reason for the electron attachment energy of fluorine being less negative than that for chlorine is because of the more tightly bound 2p electrons, leading to more interelectronic repulsion than in the chlorine atom, the repulsions not being restricted to those between electron pairs. This effect also applies to the F–F and O–O single bond lengths.

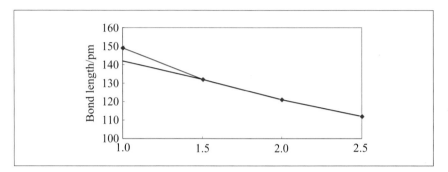

Figure 4.5 A plot of bond length against bond order for some dioxygen species

4.3   Some Heteronuclear Diatomic Molecules

The **heteronuclear** diatomic molecules nitrogen monoxide, NO, carbon monoxide, CO, and hydrogen fluoride, HF, are dealt with in this section. They belong to the point group $C_{\infty v}$ and possess a C_{∞} axis of symmetry and an infinite number of vertical planes, all containing that axis. The orbitals of the molecules NO and CO are similar to those of the E_2 molecules of the previous section, but have different labels because of their different point group. In addition to the different symmetry there are effects due to the participating atoms having different electronegativity coefficients, the energies of the combining atomic levels not being identical in such cases.

4.3.1 Nitrogen Monoxide (Nitric Oxide), NO

The molecular orbital diagram for the nitrogen monoxide molecule is shown in Figure 4.6. The orbitals are produced from the same pairs of atomic orbitals as in the cases of the homonuclear diatomic molecules of Section 4.2.

To take into account the energy inequalities between the participating atomic orbitals the wave functions have extra c terms, as in the example:

$$\phi(3\sigma^+) = c_1\psi(2p_z)_A + c_2\psi(2p_z)_B \qquad (4.10)$$

This is appropriate for the combination of the $2p_z$ orbitals of two dif-

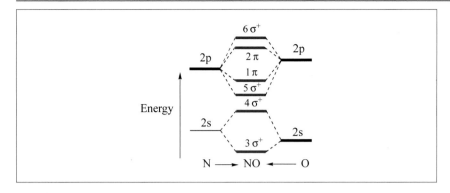

Figure 4.6 Molecular orbital diagram for nitrogen monoxide, NO

ferent atoms with different energies. The value of $(c_1)^2$ represents the contribution of the $\psi(2p_z)$ wave function of atom A to the molecular orbital $\phi(3\sigma^+)$; likewise, the value of $(c_2)^2$ represents the contribution of the $\psi(2p_z)$ wave function of atom B.

The first ionization energies of nitrogen (1400 kJ mol^{-1}) and oxygen (1314 kJ mol^{-1}) are quite similar and the 2p–2s energy gap is smaller in N than in O, so the atomic orbitals match up reasonably well. The symmetry terminology is different, and because of the absence of an inversion centre there are no g or u subscripts and this alters the numbering of the σ and π orbitals. The two pairs of 1s electrons (or KK) form the $1\sigma^+$ and $2\sigma^+$ orbitals; the other σ^+ orbitals follow in order of increasing energy. The two sets of π orbitals become 1π (bonding) and 2π (antibonding), respectively. The electronic configuration of the nitrogen monoxide molecule is thus: $KK(3\sigma^+)^2(4\sigma^+)^2(1\pi)^4(5\sigma^+)^2(2\pi)^1$. The bond order is 2.5, consistent with the bond dissociation energy (626 kJ mol^{-1}) and bond length (114 pm). The molecule is chemically very reactive and is paramagnetic because of the single unpaired anti-bonding electron. If this electron is removed to give the positively charged NO$^+$ ion the bond order increases to 3.0. The bond length in the NO$^+$ ion is 106 pm. The first ionization energy of NO, from its photoelectron spectrum, is 897 kJ mol^{-1}. The vibrational spacing in the NO$^+$ ion is 27 kJ mol^{-1}, compared with that of 22.6 kJ mol^{-1} in the NO molecule, evidence that the first ionization is from an anti-bonding orbital.

In the NO case (ignoring the 1s^2 pairs) there are four orbitals possessing the same symmetry, σ^+, which can therefore mix to an extent which depends upon energy differences. It would be expected that all four orbitals would possess 2s and 2p contributions, although it is not possible to quantify those in qualitative molecular orbital theory. The 2p–2s gaps in nitrogen and oxygen are relatively large, so that the 2p–2s mixtures in the σ^+ molecular orbitals would not be expected to approach equivalence.

The NO molecule, because of its single unpaired electron, is sometimes written as NO•, the 'dot' signifying the 'odd' electron. This has given rise in the biochemical literature to some serious misunderstandings, particularly in the rôle of NO as a 'chemical messenger,' with the formulae NO and NO• being ascribed to species having different chemical properties!

Worked Problem

Q Which of the molecules N_2 or NO would you expect to have the larger first ionization energy? State your reasoning.

A N_2 would be expected to have a larger first ionization energy than NO since the latter molecule has an anti-bonding electron that is easily removed.

4.3.2 Carbon Monoxide

As the first ionization energies imply, the energies of the atomic orbitals of carbon (first ionization energy = 1086 kJ mol^{-1}) and oxygen (first ionization energy = 1314 kJ mol^{-1}) do not match up very well. When the 2p–2s energy gaps are compared (C, 386, and O, 1544 kJ mol^{-1}) it is obvious that there is a great mismatch between the respective 2s levels. That of the oxygen atom is too low to interact with the 2s atomic orbital of the carbon atom in any significant manner. The 2s atomic orbital of the oxygen atom may be regarded as a virtually non-bonding orbital ($3\sigma^+$). The small 2p–2s energy gap in the carbon atom facilitates the mixing, or **hybridization**, of its 2s and $2p_z$ orbitals, assuming that the molecular axis is coincident with the z axis. The two orbitals participate in σ^+ orbitals of the molecule and can mix. Because of the relatively small 2p–2s energy gap they do mix, the new orbitals being written as:

$$h_1(C) = 1/2^{1/2}[\psi(2s)_C + \psi(2p_z)_C] \tag{4.11}$$

and

$$h_2(C) = 1/2^{1/2}[\psi(2s)_C - \psi(2p_z)_C] \tag{4.12}$$

The normalization factors are included in equations 4.11 and 4.12 and the minus sign in equation 4.12 means 'combine the 2s orbital with the 2p orbital after it has been reversed so that its positive lobe is directed in the opposite direction from that in equation 4.11'.

The hybrid orbitals, h_1 and h_2, are shown diagrammatically in Figure 4.7, together with the atomic orbitals from which they are formed. They have two unequal lobes of oppositely signed ψ values, the large positive lobe of h_1 being directed at the oxygen atom, with that of h_2 pointing in the opposite direction. It would be difficult for h_2 to play any significant part in the bonding to the oxygen atom, and it should be regarded as non-bonding (it is labelled $5\sigma^+$ in Figure 4.8).

The interaction of $h_1(C)$ and the $2p_z$ orbital of the oxygen atom gives a bonding combination:

$$\phi(4\sigma^+) = h_1(C) + 2p_z(O) \tag{4.13}$$

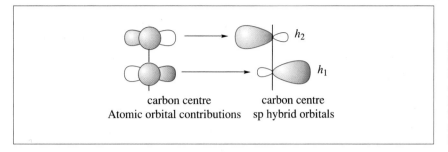

Figure 4.7 The formation of the two sp hybrid orbitals of the carbon atom used in the production of molecular orbitals in the CO molecule. h_1 is directed towards the oxygen atom, h_2 is directed diametrically opposite to the CO bonding region

and an anti-bonding combination:

$$\phi(6\sigma^+) = h_1(C) - 2p_z(O) \tag{4.14}$$

The molecular orbitals are completed by the interaction of the two sets of $2p_x$ and $2p_y$ orbitals to give doubly degenerate 1π (bonding) and 2π (anti-bonding) levels:

$$\phi(1\pi) = \psi(2p_{x,y})_C + \psi(2p_{x,y})_O \tag{4.15}$$

$$\phi(2\pi) = \psi(2p_{x,y})_C - \psi(2p_{x,y})_O \tag{4.16}$$

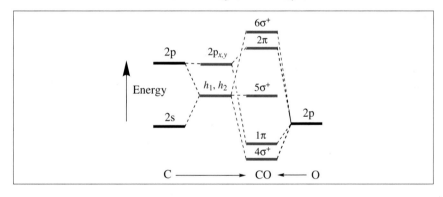

Figure 4.8 A molecular orbital diagram for the CO molecule

The molecular orbital diagram is given in Figure 4.8. The electronic configuration of the CO molecule is thus $KK(3\sigma^+)^2(4\sigma^+)^2(3\pi)^4(5\sigma^+)^2$ and the bond order is 3, consistent with the high bond dissociation energy of 1090 kJ mol^{-1} and the short bond length of 113 pm. Carbon monoxide is a relatively inert chemical substance, but it does have an extensive involvement with the lower oxidation states of the transition elements with which it forms a great many carbonyl complexes, in which it acts as a ligand. The bonding of CO to a transition metal involves the use of the otherwise non-bonding electron pair in the $5\sigma^+$ orbital. The vacant 2π orbital is also important in the bonding of CO to transition metals, a subject not treated in this book.

The cyanide ion, CN^-, is isoelectronic with carbon monoxide and has an extensive chemistry of reaction with transition metals (*e.g.* the formation of the hexacyanoferrate(III) ion, $[Fe(CN)_6^{3-}]$ by reaction with iron(III) in solution) but, unlike CO, it shows a preference for the positive oxidation states of the elements. This is mainly because of its negative charge.

Worked Problem

Q Estimate the bond order for the molecule CN, and compare the value with that for the CN^- ion. Should the bond in CN be shorter or longer than that in CN^-?

A The molecular orbital diagram for CO (Figure 4.8) may be used to solve this problem. The CN^- ion is *isoelectronic* with CO and thus has the same electronic configuration, $KK(3\sigma^+)^2(4\sigma^+)^2(3\pi)^4$ $(5\sigma^+)^2$, giving a bond order of 3. To form the neutral molecule CN an electron is removed from the $5\sigma^+$ bonding orbital, so the bond order is reduced to 2.5. The bond in CN should be longer than that in the CN^- ion.

4.3.3 The Photoelectron Spectra of N_2 and CO

The photoelectron spectra of the isoelectronic molecules N_2 and CO have peaks corresponding to the ionization energies, I, given in Table 4.3. The peaks corresponding to the ionized molecules have vibrational structure, with vibrational separations in wavenumbers given in Table 4.3. The fundamental vibration frequencies of the neutral molecules N_2 and CO are 2345 and 2143 cm^{-1}, respectively.

Table 4.3 Photoelectron spectra data for N_2 and CO

N_2 I/kJ mol^{-1}	v/cm^{-1}	CO I/kJ mol^{-1}	v/cm^{-1}
1503	2150	1352	2184
1639	1810	1632	1535
1812	2390	1903	1678

The data in Table 4.3 allow a discussion of the bonding/anti-bonding characteristics of the three orbitals from which electrons are removed in the three ionization processes for both molecules, and the relationship of these characteristics to the formal bond order in the neutral molecules. A comparison of the vibrational separation in the neutral molecule with that in an ion produced by photoionization should indicate whether the electron removed originated in a bonding, non-bonding or anti-bonding orbital. In practice the decision is not straightforward, because of the interactions that may occur between orbitals of the same symmetry. In the N_2 case [electronic configuration $KK(2\sigma_g^+)^2(2\sigma_u^+)^2$ $(1\pi_u)^4(3\sigma_g^+)^2$] the first ionization is from the $3\sigma_g^+$ orbital, which is formally a bonding orbital. The vibrational separation of the N_2^+ ion [$(2\sigma_g^+)^2(2\sigma_u^+)^2(1\pi_u^4)(3\sigma_g^+)^1$] is smaller than that of the neutral molecule, but not by the amount possibly expected for the removal of a bonding electron. From the evidence, the $3\sigma_g^+$ orbital can be described as weakly bonding. Its relative instability arises because of the interaction with the low-energy $2\sigma_g^+$ orbital, which gains stability thereby. The second ionization is from the highly bonding $1\pi_u$ level, the vibrational separation in the N_2^+ ion [$(2\sigma_g^+)^2(2\sigma_u^+)^2(1\pi_u^3)(3\sigma_g^+)^2$] showing a significant fall compared to the neutral molecule. The third ionization is from the ostensibly anti-bonding $2\sigma_u^+$ orbital, but the evidence from the vibrational separations of N_2^+ [$(2\sigma_g^+)^2(2\sigma_u^+)^1(1\pi_u)^4(3\sigma_g^+)^2$] is that it is practically non-bonding. This is because of the interaction with the higher energy vacant anti-bonding $3\sigma_u^+$ orbital, the $2\sigma_u^+$ orbital gaining some stabilization at the expense of the vacant orbital. So, in the N_2 molecule with its formal bond order of 3, it may be concluded that this value is due to two pairs of bonding π electrons plus a combination from a not-very-bonding $3\sigma_g^+$ pair of electrons and a pair of $2\sigma_u^+$ electrons which, rather than being anti-bonding, are virtually non-bonding. Considerations such as these put into perspective the somewhat over-simplified conclusions that may be made from an overall bond order about the state of bonding in a molecule.

The first ionization of the CO molecule is from a virtually non-bonding orbital to give the CO^+ ion [$(4\sigma^+)^2(1\pi)^4(5\sigma^+)^1$]; the second and third ionizations are from strongly bonding molecular orbitals to give the [$(4\sigma^+)^2(1\pi)^3(5\sigma^+)^2$] and [$(4\sigma^+)^1(1\pi)^4(5\sigma^+)^2$] CO^+ ions. Thus, in the CO case the connection between bond order and the energies of the contributing orbitals is much clearer than in the N_2 case. The non-bonding $5\sigma^+$ orbital of CO is at a higher energy than the corresponding orbital in N_2 ($3\sigma_g^+$) and is localized on the carbon atom, so that it is more easily donated to a metal atom or ion than the N_2 orbital. CO forms more stable and more numerous complexes than does N_2.

4.3.4 Hydrogen Fluoride, HF

The molecule of hydrogen fluoride, HF, belongs to the $C_{\infty v}$ point group. The hydrogen atom uses its 1s atomic orbital to make bonding and anti-bonding combinations with the $2p_z$ orbital of the fluorine atom, the z

axis coinciding with the molecular axis. Because of the different values of the first ionization energies of the two elements (hydrogen, 1312, and fluorine, 1681 kJ mol^{-1}), the molecular orbital diagram, shown in Figure 4.9, is considerably skewed, with the lower energy bonding orbital ($3\sigma^+$) having a major contribution from the fluorine $2p_z$ orbital. The higher energy anti-bonding orbital ($4\sigma^+$) has a major contribution from the hydrogen 1s orbital. Such inequality in contributing to molecular orbitals may be expressed by considering the coefficients of the atomic orbital contributions to the molecular wave functions that are constructed from linear combinations of the 1s(H) and $2p_z$(F) wave functions:

$$\phi(3\sigma^+) = c_1\psi(1s)_H + c_2\psi(2p_z)_F \qquad (4.17)$$

and

$$\phi(4\sigma^+) = c_3\psi(1s)_H - c_4\psi(2p_z)_F \qquad (4.18)$$

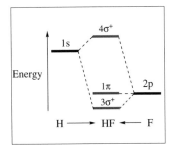

Figure 4.9 The molecular orbital diagram for the HF molecule

The **dipole moment** of a molecule is usually tabulated in terms of units called Debyes (after P. Debye, Nobel Prize for Chemistry, 1936) which, in S.I. units, are given by:

1 D = 3.336 × 10^{-30} Coulomb metre; the dipole moment of HF is 1.82 D

Because the squares of wave functions are proportional to electron probabilities, the 'share' of the two bonding electrons experienced by the fluorine atom is given by the factor $(c_2)^2/[(c_1)^2 + (c_2)^2]$. The 'share' experienced by the hydrogen atom is $(c_1)^2/[(c_1)^2 + (c_2)^2]$. The HF molecule possesses a dipole moment of 6.33 × 10^{-30} C m, which is experimental evidence for such a charge separation. The dipole moment (μ) in units of C m is given by the product of the charge (Q in coulombs) at the positive end of the molecule and the distance (in metres) between the two ends of the molecule (the bond length in the case of a diatomic molecule): $\mu = Qr$. The charge separation in HF is thus given by:

$$Q = \mu/r = (6.33 \times 10^{-30}) \div (0.092 \times 10^{-9}) = 6.88 \times 10^{-20} \text{ C}$$

which is equal to an electronic charge of $(6.88 \times 10^{-20}) \div (1.6 \times 10^{-19}) = 0.43e$. The hydrogen end of the molecule is, in effect, positive to the extent of $0.43e$, whilst the fluorine end is negative to the extent of $-0.43e$. Valence bond theory would describe the bonding as 57% covalent and 43% ionic.

The other orbitals in Figure 4.9 are the non-bonding $2\sigma^+$ (the 2s atomic orbital of the fluorine atom) and the non-bonding 1π (the $2p_x$ and $2p_y$ atomic orbitals of the fluorine atom). The $1s^2$ pair of fluorine electrons ($1\sigma^+$) is omitted from the diagram.

The discrepancy in the matching up of the 1s(H) and $2p_z$(F) orbitals is 1681 − 1312 = 369 kJ mol^{-1}, and leads to a bond with considerable ionic character. The difference in energy between the orbitals of the sodium atom (the 3s ionization energy = 496 kJ mol^{-1}) and the fluorine atom (the $2p_z$ ionization energy = 1681 kJ mol^{-1}) amounts to 1185 kJ mol^{-1}, and leads to the conclusion that covalent bond formation is impossible between Na and F. If the elements are to combine at all, it has to be by an alternative method (ionic bonding), which is described in Chapter 7.

Summary of Key Points

1. The symmetry concepts of Chapter 2 and those of molecular orbital theory were applied to the construction of molecular orbitals for a range of diatomic molecules.

2. The method of approach to a molecular problem was described.

3. The homonuclear diatomic molecules of the second period (Li_2 to Ne_2) were described in detail.

4. Some heteronuclear diatomic molecules were described in detail. Differences in electronegativity coefficients between the combining atoms were shown to be important.

5. The lengths and strengths of bonds are related to the electronic configurations of the molecules and their bond orders.

6. Where possible, the molecular orbital diagrams are related to observed photoelectronic spectra.

Problems

1. The dissociation energies, D, of the M_2 Group 1 molecules in the gas phase are:

	Li_2	Na_2	K_2	Rb_2	Cs_2
D/kJ mol^{-1}	108	73	49	47	43
$r_{covalent}$/pm	134	157	202	216	235
$r_{metallic}$/pm	152	186	227	248	265

Correlate the variation of D down the group with the covalent and metallic radii of the metal atoms.

2. Use the data given in the text to plot the O–O bond dissociation energy of the species O_2, O_2^+, O_2^- and O_2^{2-} against their bond order. On the same graph plot the O–O bond length against bond order for the same species. Specify the magnetic behaviour of each species and explain the correlations.

3. The ionization of the molecule to give the N_2^+ ion causes the bond order to be reduced to 2.5, with consequent weakening (bond dissociation energy = 841 kJ mol^{-1}) and lengthening (bond length = 112 pm) of the bond as compared to that in N_2. The N_2^- ion is produced by adding an electron to the anti-bonding $1\pi_g$ level and also has a bond order of 2.5. The bond in N_2^- is significantly weaker (bond dissociation energy = 765 kJ mol^{-1}) and longer (bond length = 119 pm) than the one in N_2. Explain these observations in terms of molecular orbital theory.

4. The HCl molecule has a dipole moment of 1.03 D and a bond length of 128 pm. Calculate the percentage ionic character, and compare your answer with that for HF. Comment on the differences in electronegativity coefficients in the two molecules.

5. The dipole moments of CO and NO are 0.1 D and 0.166 D, respectively. The oxygen atom is the positive end of the CO dipole, despite the difference in electronegativity coefficients of the two atoms which would imply the opposite conclusion. Consider the electronic configurations of the two molecules, and explain the anomalous properties of the CO molecule.

Reference

R. S. Mulliken, C. A. Rieke, D. Orloff and H. Orloff, *J. Chem. Phys.*, 1949, **17**, 1248. Formulas and numerical tables for overlap integrals.

Further Reading

D. M. P. Mingos, *Essentials of Inorganic Chemistry 1*, Oxford University Press, Oxford, 1995. This is an A–Z account of the major concepts used in modern inorganic chemistry.

5

Covalent Bonding III:
Triatomic Molecules;
Bond Angles

This chapter consists of the application of molecular orbital theory to the bonding in triatomic molecules, with its extension to the factors responsible for the determination of bond angles and molecular shapes. Additionally, a straightforward method of predicting molecular shape, the valence shell electron pair repulsion theory (VSEPR), is introduced. This is based on a valence bond approach, and is of general application. In this chapter, the discussions are restricted to: (1) the linear and bent forms of the water molecule, (2) two linear molecules, CO_2 and XeF_2, and the linear ion I_3^-, (3) the linear nitronium ion, NO_2^+, and the bent nitrogen dioxide molecule and the nitrite [or nitrate(III)] ion, NO_2^-, and (4) the linear HF_2^- ion, which is an example of strong hydrogen bonding.

Aims

By the end of this chapter you should understand:

- The applications of valence bond theory, the valence shell electron pair repulsion theory, to the bonding and bond angles of triatomic molecules
- The application of molecular orbital theory to triatomic molecules, their bond lengths and strengths
- The factors that govern bond angles
- The photoelectron spectra of selected triatomic molecules
- The nature of strong hydrogen bonding

5.1 Triatomic Molecules

Triatomic molecules may be linear or bent (*i.e.* V-shaped or angular). The shape adopted by any particular molecule is that which is consistent with the minimization of its total energy.

The **valence shell electron pair repulsion theory (VSEPR)**, is based upon the original ideas of Sigwick and Powell, and was extended by Gillespie and Nyholm. The basis of the method is that the shape of a molecule results from the minimization of the repulsions between the *pairs of electrons* in the valence shell. The term '**valence shell**' refers to the orbitals of the central atom of the molecule which could possibly be involved in bonding. The theory is operated by counting the number of electrons in the valence shell of the central atom, together with suitable contributions from the ligand atoms: one electron per bond formed or two if coordinate bonds are used. This sum is then divided by two to give the number of pairs of electrons. These electron pairs are then considered to repel each other (ignoring the interelectronic repulsion between the electrons in each pair) to give a spatial distribution which is that corresponding to the minimization of the repulsive forces. This method of predicting molecular shape affords a very rapid and easy-to-use set of arguments which almost always produces an answer which is consistent with observation.

The extension of molecular orbital theory to triatomic molecules is a major part of this chapter. It gives a very satisfactory description of the shapes and the bonding of molecules in general, and is consistent with observations of photoelectron and electronic absorption spectra. It is not possible for the VSEPR theory to explain these latter observations.

5.2   Valence Shell Electron Pair Repulsion Theory

The basis of the VSEPR theory is that the shape of a molecule (or the geometry around any particular atom connected to at least two other atoms) is assumed to be dependent upon the minimization of the repulsive forces operating between the pairs of 'sigma' (σ) valence electrons. This is an important restriction. Any 'pi' (π) or 'delta' (δ) pairs are discounted in arriving at a decision about the molecular shape. The terms 'sigma', 'pi' and 'delta' refer to the type of overlap undertaken by the contributory atomic orbitals in producing the molecular orbitals, and are referred to by their Greek-letter symbols in the remainder of the book.

Examples of the types of overlap which give rise to σ orbitals are shown in Figure 5.1. The sideways overlap of two p orbitals that gives rise to π orbitals is shown in Figure 4.1.

Figure 5.2 demonstrates the different types of overlap for d_{xz}–d_{xz} (in the xz plane) and d_{xy}–d_{xy} (along the z axis). The 'sideways' (two lobes) overlap by the d_{xz} orbitals is π-type, the four-lobe overlap by the two d_{xy} orbitals being termed δ-type. The Greek letters are used loosely for such orbitals: strictly they should be used for molecules possessing either $D_{\infty h}$ or $C_{\infty v}$ symmetry, but they are used generally to describe the type of overlap. Only σ pairs are counted up in the VSEPR approach. This often

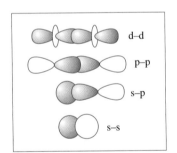

Figure 5.1 Examples of orbital overlaps which give rise to σ orbitals

means that some assumptions about the bonding have to be made before the theory may be applied.

Table 5.1 contains the basic shapes adopted by the indicated numbers of σ electron pairs, and Figure 5.3 shows representations of the basic shapes assumed by the various electron pair distributions. The **bond angles** (the angles formed by ligand–central atom–ligand triplets) are included in Table 5.1.

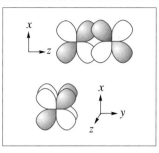

Figure 5.2 Overlap diagrams for d_{xz}–d_{xz} (in the *xz* plane) and d_{xy}–d_{xy} (along the *z* axis)

Table 5.1 The basic shapes, bond angles and hybridizations adopted by various numbers of σ pairs of electrons

Number of σ electron pairs	Distribution of electron pairs	Bond angles	Type of hybridization
2	Linear	180°	sp
3	Trigonally planar	120°	sp^2
4	Tetrahedral	109°28'	sp^3
5	Trigonally bipyramidal	180°, 120°, 90°	dsp^3
6	Octahedral	180°, 90°	d^{2}sp^3
7	Pentagonally bipyramidal	180°, 90°, 72°	d^{3}sp^3

Worked Problem

Q The valence shell of the central atom of a molecule possesses two bonding and three non-bonding pairs of electrons. State, with reasoning, the shapes of (i) the electron pair distribution and (ii) the molecule.

A There are five electron pairs in total and these would be expected to be distributed as in a trigonal bipyramid around the central atom. The three non-bonding pairs would normally occupy the trigonal plane (there is a general rule that, in the case of a trigonally pyramidal distribution of electrons, any non-bonding pairs occupy positions in the trigonal plane, dealt with in Chapter 6) so that the two ligand atoms would be as far apart as possible, leading to a linear molecule.

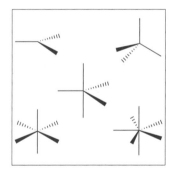

Figure 5.3 The basic shapes for the spatial distribution of three (trigonally planar), four (tetrahedral), five (trigonally bipyramidal), six (octahedral) and seven (pentagonally bipyramidal) pairs of electrons around a central atom

Table 5.1 also includes the types of **hybridization** for each of the basic shapes which are used in VSEPR and valence bond theories. Hybridization, or mixing of the indicated atomic orbitals, in each case produces hybrid orbitals that radiate from the central atom to where the ligand atoms are situated. This is done to ensure that there are localized

There are rare cases of molecules with central atoms surrounded by more than seven ligand atoms or groups. The rules of VSEPR theory apply to such cases and examples are to be found in more advanced texts.

electron-pair bonds between the central atom and the ligand atoms which constitute the molecule. Although hybridization is used in molecular orbital theory, it is only used when allowed by the symmetry rules, *i.e.* only those orbitals of identical symmetry may mix.

The application of VSEPR theory to triatomic molecules is exemplified by considering water, carbon dioxide, xenon difluoride and a trio of connected species: the nitronium ion, NO_2^+, nitrogen dioxide and the nitrite [or nitrate(III)] ion, NO_2^-.

5.2.1 Water, H_2O

The isolated water molecule has a bond angle of 104.5°. The central oxygen atom has the valence shell electron configuration $2s^2 2p^4$, with two of the 2p orbitals singly occupied. The hydrogen atoms supply one electron each to the valence shell of the central atom, which makes a total of eight σ electrons (since the 1s–2p overlaps are σ-type). The four electron pairs are most stable when their distribution is tetrahedral. Two of the pairs are bonding pairs, the other two being non-bonding or lone pairs. The production of the four pairs of electrons in the valence shell of the oxygen atom is shown in Figure 5.4. The box notation is useful for VSEPR theory, and Figure 5.4 shows the valence state of the oxygen atom with its two vacancies in the 2p set of atomic orbitals ready to accept the two electrons from the hydrogen atoms when the water molecule is formed. The in-coming electrons are shown in red; this is done to help with the electron counting rather than to suggest that electrons can be distinguished from each other.

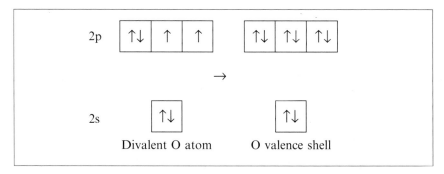

Figure 5.4 A diagram showing the two-valent state of the oxygen atom and the use of the oxygen valence shell to accommodate the electrons from the hydrogen atoms in the formation of a water molecule; the electrons shown in red are those from the ligand hydrogen atoms

This VSEPR picture of the water molecule implies that the bond angle should be that of the regular tetrahedron, 109°28′. The theory predicts that the water molecule should be bent, which it is. To refine the prediction of the bond angle it is possible to argue that the non-bonding pairs of electrons, which are more localized on the central oxygen atom than the bonding pairs, have a greater repulsive effect upon the bonding pairs than that exerted by the bonding pairs upon each other. The result would be a reduction of the angle between the two bonding pairs, making the bond angle smaller than the regular tetrahedral value. It is not possible to quantify this aspect of the theory.

5.2.2 Carbon Dioxide, CO_2

The carbon dioxide molecule is linear and belongs to the $D_{\infty h}$ point group. The carbon atom has the valence shell configuration of $2s^2 2p^2$. In that state the carbon atom is **divalent**: there are two unpaired electrons which occupy two of the three 2p atomic orbitals. The two singly occupied orbitals may each accept an electron from a ligand atom. In order for carbon to be **tetravalent** it is necessary to arrange for there to be four unpaired electrons in its valence shell. This may be achieved (as a thought experiment) by transferring one of the 2s electrons to the previously unoccupied 2p orbital. It is then possible to feed two σ electrons into the 2s and one of the 2p orbitals, leaving the other two 2p orbitals to accept the two π electrons. These hypothetical electronic changes are shown in Figure 5.5.

In general, it is assumed in VSEPR theory that the repulsive interactions between bonding (b) and non-bonding or lone (l) electron pairs are ordered:

$$b \leftrightarrow b < b \leftrightarrow l < l \leftrightarrow l$$

the logic of this being that lone pairs are localized on the central atom and are a more concentrated source of charge than bonding pairs which are shared between two atoms and are a less concentrated source.

It is common practice, in VSEPR thought experiments in which electrons are unpaired and/or excited to higher energy orbitals, to reverse electron spins. Spin reversal is a forbidden transition in spectroscopy, but is of no consequence in electron pair construction in valence shells.

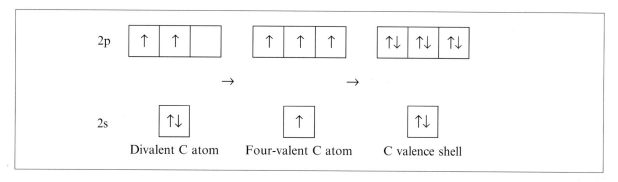

Prior knowledge is used of the normal divalency of the oxygen atom and the impossibility of there being more than one σ bond between any pair of atoms (the spatial distribution of atomic orbitals decides that). The count of σ electron pairs comes to two and quite correctly the theory predicts the shape of the CO_2 molecule to be linear, the two σ pairs repelling each other to a position of minimum repulsion. The two π pairs are used to complete the bonding picture. In valence bond theory the description is taken further by having the s and one of the p orbitals undertake hybridization to give two sp hybrid orbitals diametrically opposed to each other to give the molecule its linearity, with the two pure p orbitals reserved for the two π pairs.

Figure 5.5 A diagram showing the electronic changes made in the formation of CO_2; the electrons shown in red are those from the ligand oxygen atoms

5.2.3 Xenon Difluoride, XeF$_2$, and the Octet Rule

The XeF$_2$ molecule is linear. According to the octet rule, which is assumed to apply to the central atoms of most main group compounds and which states that such central atoms combine with other atoms until they have a share in eight electrons, *i.e.* an s^{2}p^6 configuration is produced, the xenon atom should be zerovalent.

Box 5.1 Compounds of Xenon

Compounds of xenon were first prepared in 1962, following Neil Bartlett's observation that dioxygen could be oxidized with PtF$_6$ to give the compound O$_2^+$[PtF$_6^-$] and that the first ionization energies of dioxygen and xenon were almost identical (1165 and 1170 kJ mol^{-1}, respectively). The compound formed by heating xenon with PtF$_6$ was shown to be [XeF$^+$][Pt$_2$F$_{11}^-$], although it was initially thought to have the formula Xe$^+$PtF$_6^-$. The XeF$^+$ cation is isoelectronic with IF. The next compounds of xenon to be discovered were the three fluorides, XeF$_2$, XeF$_4$ and XeF$_6$, made by passing mixtures of xenon and fluorine through a heated nickel tube. The predominance of the octet theory prevented such a simple experiment being attempted for over 60 years, for both elements were available in 1898! Exceptions to the octet rule abound in the higher oxidation states of the elements of Groups 15, 16 and 17, *e.g.* PCl$_5$, SF$_6$ and IF$_7$, in which there are 10, 12 and 14 electrons, respectively, in the valence shells of the P, S and I atoms. These compounds are assumed to have d orbital participation in the VSEPR treatments of their bonding.

VSEPR theory acknowledges that the valence shell of the xenon atom has the configuration 5s^{2}5p^6, and as such the atom should be zerovalent. There is the possibility of causing a 5p to 5d excitation to make the atom divalent. Addition of the two valence electrons from the fluorine atoms would make two bonding pairs and these, together with the remaining three pairs of non-bonding electrons, would contribute to the total of five electron pairs. These electronic changes are shown in Figure 5.6.

The 5d orbital used would be the d$_{z^2}$, which would participate in hybridization with the 5s and three 5p orbitals to give five dsp^3 hybrid orbitals disposed towards the vertices of a trigonal bipyramid. The five pairs would assume a trigonally bipyramidal distribution, and it is logical for the fluorine atoms to be as far away from each other as possible to give a linear molecule. The alternative approach would divide the

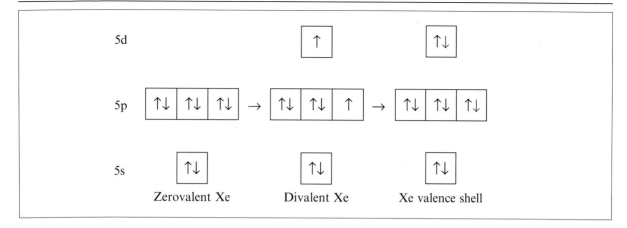

molecule up into Xe^{2+} and two ligand fluoride ions. The Xe^{2+} ($5s^25p^4$) would then accept two electron pairs from the fluorine atoms, one of which would fill the third 5p orbital and the second would enter the next available level, 5d. The same (correct) conclusion would follow.

Figure 5.6 A diagram showing the electronic changes made in the formation of XeF_2; the electrons shown in red are those from the ligand fluorine atoms

5.2.4 The Tri-iodide ion, I_3^-

The tri-iodide ion, I_3^-, is linear. VSEPR theory may be applied by assuming the **coordinate bond** structure $I^-\rightarrow I^+\leftarrow I^-$, in which the central atom has an **oxidation state** of +1 and acts as the acceptor of **electron pairs** from the two ligand iodide ions. The electronic configuration of iodine in its valence shell is $5s^25p^5$, and ionization to I^+ removes one of the 5p electrons to give the configuration $5s^25p^4$. If the 5p electrons are arranged to be paired up, there is a vacant 5p orbital which can accept one of the incoming electron pairs from the ligand iodide ions.

The tri-iodide ion is responsible for the solubility of elemental iodine in an aqueous solution of potassium iodide and gives rise to the deep yellow-brown colour of the solution, unlike the violet colour displayed by the iodine molecule in the vapour phase or in a solution of ethanol:

$$I^-(aq) + I_2(s) \rightarrow I_3^-(aq)$$

Box 5.2 Oxidation State and Valency

These terms are frequently used and may lead to confusion, if used in the wrong context. As a reminder, the valency of an atom is strictly the number of bonds (including σ, π and δ) in which it participates in any particular compound. In IF_7, the iodine atom participates in seven σ bonds to the fluorine atoms. The fluorine atoms are individually monovalent. Oxidation states are more useful than valency in describing ionic compounds. In the crystalline solid CaF_2, the calcium is best thought of as composed of calcium dications and fluoride anions, $Ca^{2+}(F^-)_2$. The calcium is in its +2 oxidation state, having lost its valency electrons, and the fluorine atoms are in the –1 oxidation state, both having accepted an electron from the calcium atom. The iodine atom in IF_7 may be thought of as being in its +7 oxidation state, if the fluorine atoms are in their –1 oxidation state. This would be indicated by a Roman numeral superscript on the element symbol, *e.g.* I^{VII}.

The second donated electron pair must then occupy one of the 5d orbitals (this would be the $5d_{z^2}$ orbital, assuming that the tri-iodide ion is coincident with the Cartesian *z* axis), giving five σ electron pairs which would be directed towards the vertices of a trigonal bipyramid. The three non-bonding pairs occupy the trigonal plane, so that the ligand iodine atoms are placed in the two axial (or apical) positions, making the tri-iodide ion linear.

The alternative approach is to place the negative charge on the central iodine atom so that it has the xenon configuration, $5s^2 5p^6$. The divalent state of I^- is produced by promoting one of the 5p electrons to a 5d orbital and this results in the same conclusion from the two bonding pairs and three non-bonding pairs as the previous treatment.

Valence bond theory resorts to using the two forms of the tri-iodide ion as canonical forms contributing to the resonance hybrid:

$$I^- \rightarrow I^+ \leftarrow I^- \leftrightarrow I-I^--I$$

5.2.5 Hypervalence

The exceptions to the octet rule described in the previous section, the xenon compounds and the tri-iodide ion, are dealt with by the VSEPR and valence bond theories by assuming that the lowest energy available d orbitals participate in the bonding. This occurs for all main group compounds in which the central atom forms more than four formal covalent bonds, and is collectively known as **hypervalence**, resulting from the *expansion of the valence shell*. This is referred to in later sections of the book, and the molecular orbital approach is compared with the valence bond theory to show that d orbital participation is unnecessary in some cases. It is essential to note that d orbital participation in bonding of the central atom is dependent upon the symmetry properties of individual compounds and the d orbitals.

Worked Problem

Q Give examples of species from Groups 15, 16, 17 and 18 which exhibit hypervalence. Explain why oxygen and fluorine do not participate in hypervalency.

A Phosphorus, sulfur, chlorine and xenon atoms would be expected to have valencies of 3, 2, 1 and 0 respectively if they did not show hypervalence. Compounds of the elements which exhibit hypervalence are PCl_5, SF_6, ClF_7 and XeF_4. Oxygen and fluorine do not participate in hypervalency because they do not have any low-energy d orbitals which could be used to form hypervalent compounds. They do possess 3d orbitals, but these are of too high an energy to be used in compound formation.

In accord with the octet rule, the valencies or oxidation states of the main group elements depend upon their $s^{1 \ or \ 2}p^{1 \ to \ 6}$ electronic configurations. Elements of Groups 1, 2, 13 and 14 usually have one valency or oxidation state, which corresponds to the use of all the s and p valence electrons that they possess. For example, in the third period there are the species Na^+, Mg^{2+}, Al^{3+} and $Si^{IV}O_2$ in the main groups. In the case of the heavier members of Group 14, the '**inert pair effect**' operates to produce compounds with valencies or oxidation states which are two units fewer than the maximum expected, *e.g.* Sn^{2+} and PbO. The heavier elements of Groups 15 and 16 are also subject to the inert pair effect and have compounds based upon the retention of the s^2 pairs. The 'normal' valencies/oxidation states of the elements of Groups 15 to 18 are based upon the number of vacancies in the p orbitals of their atoms, resulting in the values 3, 2, 1 and 0, respectively. The range of valencies/oxidation states exhibited by compounds of Groups 15–18 is given in Table 5.2

Table 5.2 Oxidation states for elements in Groups 15–18

Group	'Normal' oxidation state	Other oxidation states
15	3	5
16	2	4, 6 (except O)
17	1	3, 5, 7 (except F)
18	0	2, 4, 6, 8 (xenon only)[a]

[a]Krypton is known to form compounds in its +2 state, and radon probably has a chemistry similar to that of xenon

Compounds in which the valencies/oxidation states of the central element are greater than the 'normal' ones, based upon the octet rule, are examples of hypervalency, and vary in steps of two because they are related to the unpairing and use of two extra electrons at each stage.

5.2.6 The Nitronium Ion, NO_2^+, Nitrogen Dioxide and the Nitrite [or Nitrate(III)] Ion, NO_2^-

The Nitronium Ion, NO_2^+

The nitronium ion may be studied in compounds such as $NO_2^+ClO_4^-$, which is prepared by dissolving N_2O_5 in perchloric acid [chloric(VII) acid]:

$$N_2O_5 + HClO_4 \rightarrow NO_2^+ClO_4^- + HNO_3$$

The best approach to the VSEPR treatment of positive ions is to remove an electron from the central atom as a first step. With the nitronium ion this results in a central N^+ ion which has the same electronic configuration as the carbon atom, $2s^22p^2$, and is divalent. A notional promotion of one of the 2s electrons into the otherwise vacant 2p orbital produces the tetravalent state $2s^12p^3$; this can accept the four electrons from the two ligand oxygen atoms, which ensures that oxygen is exerting its usual divalency. With the assumption that the N–O bonds will be a double σ,π combination, two of the four bonding pairs of the $2s^22p^6$ system are reserved for the π bonds, leaving only two bonding pairs to undergo the VSEPR bond pair–bond pair repulsion. This results in the prediction that the nitronium ion should be linear, which is found to be the case.

This is not surprising, since the ion is isoelectronic with CO_2, *i.e.* they possess the same number of electrons, and isoelectronic species normally have the same shape.

Nitrogen Dioxide

In the NO_2 molecule the central nitrogen ($2s^22p^3$) is trivalent, and to cope with the normal valencies of the oxygen atoms it is best to regard the central atom as N^+ which can then bond to O^- (isoelectronic with F) by a σ bond and to an oxygen atom by the usual double σ,π combination. By this rather artificial procedure the valence shell of the central N^+ ion is $2s^12p^3$ in a four-valent state. If two σ bonds and one π bond are made by adding three electrons, one from O^- and two from O, the valence shell is then $2s^22p^5$, which has the odd electron in one of the 2p orbitals. In VSEPR terms there are two and a half σ electron pairs, which will result in a planar distribution with the angles between the bonding pairs and the bonding half-pair being smaller than the angle between the two bonding pairs, resulting in a molecule with a bond angle somewhat greater than the 120° that would arise from a regular trigonally planar molecule. The observed bond angle in NO_2 is 134°.

Box 5.3 'Awkward' Cases in VSEPR Theory

An alternative way to deal with rather awkward cases in VSEPR theory is to take note of the oxidation state of the central ion and strip the appropriate number of electrons from the central atom, then feed in electron pairs from each of the ligand ions. In other words, treat them as ionic. In the case of NO_2 the nitrogen atom has an oxidation state of +4 and the electronic configuration of N^{4+} is $2s^1$. If the two O^{2-} ions are then assumed to bond to the N^{4+} by each supplying a pair of electrons in a dative or coordinate bond formation, the valence shell becomes $2s^1 2p^2 2p^2$ and the two and a half pairs give the required result. The nitrite ion may be regarded as the combination $N^{3+} + 2O^{2-}$, so the valence shell of the N^{3+}, $2s^2$, becomes $2s^2 2p^2 2p^2$ when the two coordinate bonds are formed: $O^{2-}{:}{\rightarrow}N^{3+}{\leftarrow}{:}O^{2-}$. This results in the observed bent ion, with an angle somewhat lower than the regular 120°. These examples show that the VSEPR theory can, in some cases, be very misleading about the bonding which occurs.

The Nitrite Ion, NO_2^-

The treatment of the nitrite ion is similar to that of NO_2, except that the artifice of having a charge separation is not necessary. The central nitrogen atom has the trivalent $2s^2 2p^3$ configuration, and can accept one electron from an O^- and two electrons from an oxygen atom to give the filled $2s^2 2p^6$ configuration. With one pair of electrons reserved for the π bond there are then three pairs of σ electrons which determine the shape of the ion. The trigonally planar distribution of the σ electron pairs produces a bent nitrite ion with the bond angle based on 120°, but with the lone pair exerting slightly more repulsion to force the bond angle to somewhat less than the regular angle. The observed angle is 115°.

The valence bond treatment of the nitrite ion includes the two canonical forms shown in Figure 5.7, which contribute to the resonance hybrid that represents the bonding. It is a method of indicating that the two bonds are equivalent in length and strength and that the bond order is 1.5 per bond, and that the negative charge is spread evenly between the two oxygen atoms.

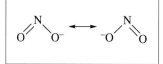

Figure 5.7 Two canonical forms of the nitrite ion

Worked Problem

Q Use the description of the operation of VSEPR theory given above for NO_2^- to draw the thought experiment box diagram which outlines the logic necessary for deciding its structure.

A (1) NO_2^-. Using the constitution $N + O + O^-$ for the ion, the box diagram shows (i) the ground state of N and (ii) the addition of two electrons from the oxygen ligand, forming $\sigma + \pi$ pairs and the addition of one electron from O^- to give another σ pair. Of the four electron pairs, two are σ bonding, one is non-bonding and the fourth is reserved for the π bond. The bond angle is then based on the three σ pairs and would be expected to be somewhat less than the regular 120°.

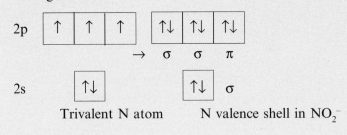

2p

→   σ   σ   π

2s

σ

Trivalent N atom   N valence shell in NO_2^-

5.3   Molecular Orbital Theory

The application of molecular orbital (MO) theory, even on a qualitative basis as is described in this book, is a more lengthy procedure than that using VSEPR theory. It begins with the atomic orbitals which are available for bonding and the number of electrons which have to be accommodated. The understanding of the bonding of a molecule arises from the proper application of symmetry theory as is demonstrated in the following sections, using the same examples that are subjected to the VSEPR and valence bond theories in previous sections.

5.3.1   Water

The groundwork for the application of MO theory to the water molecule has been carried out to a large extent in Section 2.2.1. The procedure is to identify the point group to which the molecule belongs. To demonstrate the power of MO theory, both of the extreme geometries of the molecule, the bent (bond angle, 90°) and linear (bond angle, 180°) forms, are treated.

90° Water Molecule

The general protocol for the derivation of the MO energy diagram for any molecule is outlined in this section. It is applied to the 90° water molecule as follows.

1. Identify the point group to which the molecule belongs.
The 90° form of the water molecule belongs to the C_{2v} point group. There

is only one axis of symmetry, C_2, and this is arranged to coincide with the z axis. The position of the molecule with respect to the coordinate axes is as shown in Figure 5.8.

2. Classify the atomic orbitals in the valency shell of the central atom with respect to the point group of the molecule.

The classification of the orbitals of the oxygen atom is a matter of looking them up in the C_{2v} character table, a full version of which is included in Appendix 1. The 2s(O) orbital transforms as an a_1 representation, the $2p_x$(O) orbital transforms as a b_1 representation, the $2p_y$(O) orbital transforms as a b_2 representation and the $2p_z$(O) orbital transforms as another a_1 representation.

3. Classify the valency orbitals of the ligand atoms with respect to the point group of the molecule and identify their group orbitals.

The two hydrogen 1s orbitals have the characters shown in the following representation; the individual characters are the number of orbitals *unaffected* by the particular symmetry operation carried out upon the two orbitals.

C_{2v}	E	C_2	$\sigma_v(xz)$	$\sigma_v'(yz)$
1s + 1s	2	0	0	2

This representation may be seen, by inspection of the C_{2v} character table, to be equivalent to the sum of the characters for the a_1 and b_2 irreducible representations. Those particular hydrogen group orbitals are very similar to the ones used to describe the bonding of the H_2 molecule (equations 3.3 and 3.4, respectively, and shown in Figure 3.1), but the nuclei are further apart in water. The a_1 hydrogen group orbital (labelled as h_1) is H–H bonding, whereas the b_2 group orbital (labelled as h_2) is H–H anti-bonding. Because of the greater distance between the hydrogen atoms in the water molecule, the bonding and anti-bonding interactions are not as great as those in the H_2 molecule.

4. Correlate the orbitals of the central atom with the group orbitals of the ligand atoms and draw the MO diagram for the molecule.

Two considerations are of importance in drawing the molecular orbital diagram for a molecule. First, the relative energies of the orbitals must be considered. The first ionization energies of the atoms involved in molecule formation give a good indication of how to position the atomic levels, and information about the magnitude of p–s energy gaps is useful. Second, and highly important, is the guiding principle that *only atomic orbitals (single or group) belonging to the same irreducible representation may combine to give bonding and anti-bonding MOs*. It is also helpful to

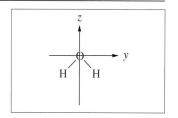

Figure 5.8 The position of the water molecule with respect to the coordinate axes; the molecule is in the *xz* plane

It is always a good exercise to check that the orbital to be classified does transform as indicated by the individual characters in a representation. This should be done with the oxygen orbitals.

have knowledge of the photoelectron and absorption spectra to assist with the exact placement of the molecular levels, although such information is unnecessary for the production of a qualitative molecular orbital diagram.

When considering the formation of the water molecule, an important observation is that the first ionization energies of the hydrogen and oxygen atoms are almost identical (H, 1312; O, 1314 kJ mol^{-1}), and there is a large energy difference between the 2p and 2s levels of the oxygen atom (1544 kJ mol^{-1}). The latter piece of information indicates that the 2s(O) orbital does not participate in the bonding to a major extent. The 2s(O) and 2p$_z$ orbitals do have identical symmetry properties and so have the possibility of *mixing* to give two **hybrid orbitals**, but the large energy gap between them prevents very much interaction. This means that the main a$_1$ combination is between the 2p$_z$ and h_1 orbitals. Combination occurs between the b$_2$ [2p$_y$(O)] and h_2 orbitals, with the b$_1$ [2p$_x$(O)] orbital remaining as a non-bonding MO. The a$_1$ interaction produces a bonding orbital which is O–H bonding and which is also H–H bonding. The bonding orbital from the b$_2$ combination is O–H bonding but is H–H anti-bonding.

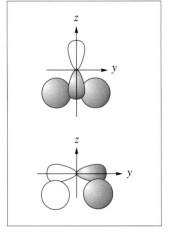

Figure 5.9 Overlap diagrams for the a$_1$ (*top*) and b$_2$ (*bottom*) bonding orbitals of the C_{2v} water molecule

The 'tie lines' between the atomic and molecular orbitals in Figure 5.10 (and all other MO diagrams in this text) do not take the mixing of MOs into account. The MO diagrams in this book are simplified as far as possible and, wherever there is orbital mixing, the contributions from the various atomic orbitals are not always fully indicated by the tie lines. Whenever orbital mixing is important, it is referred to in the text.

Worked Problem

Q In the bent water molecule, why is the 1b$_1$ orbital non-bonding?

A Because the hydrogen group orbitals have the symmetries a$_1$ and b$_2$ and, therefore, cannot interact with an orbital of any other symmetry.

The above interactions are made obvious in the pictorial representation of the a$_1$ and b$_2$ orbital combination diagrams shown in Figure 5.9, and the MO energy diagram for the 90° water molecule is shown in Figure 5.10.

The 2a$_1$ orbital gains some stability from a small interaction with the bonding 3a$_1$ orbital, which becomes destabilized to a similar extent. This causes the 3a$_1$ orbital to have a higher energy than that of the 1b$_2$ orbital.

One very important difference between VSEPR theory and MO theory should be noted. The MOs of the water molecule which participate in the bonding are three-centre orbitals. They are associated with all three atoms of the molecule. There are no *localized electron pair bonds* between pairs of atoms as used in the application of VSEPR theory. The existence of three-centre orbitals (and multi-centre orbitals in more complicated molecules) is not only more consistent with symmetry theory, it

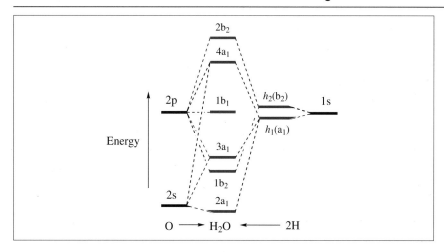

Figure 5.10 MO diagram for the 90° water molecule

also allows for the reduction of interelectronic repulsion effects when more extensive, non-localized (**delocalized**), orbitals are doubly occupied.

180° Water Molecule

1. The linear water molecule belongs to the $D_{\infty h}$ point group. The classifications of atomic and group orbitals must be carried out using the $D_{\infty h}$ character table. The molecular axis, C_{∞}, is arranged to coincide with the z axis.

2. The classification of the 2s and 2p atomic orbitals of the central oxygen atom. The 2s(O) orbital transforms as σ_g^+, the $2p_x$ and $2p_y$ orbitals transform as the doubly degenerate π_u representation, and the $2p_z$ orbital transforms as σ_u^+. In some texts the '+' and '−' superscripts are omitted, but throughout this one there is strict adherence to the use of the full symbols for all orbital symmetry representations. Unlike the 90° case, the 2s and $2p_z$ orbitals have different symmetry properties so there is no question of their mixing.

3. Classification of the hydrogen group orbitals. The H–H bonding group orbital, h_1, transforms as σ_g^+, and the H–H anti-bonding group orbital, h_2, transforms as σ_u^+. The detailed conclusions are dealt with in Section 3.2.1.

4. The MOs may now be constructed by allowing the orbitals of the oxygen and hydrogen atoms, belonging to the same representations, to combine. Thus the interaction between σ_g^+ and h_1 is possible, by symmetry, but is very restricted in extent by the large difference in energy between the two orbitals. The major interaction is between $\sigma_u^+(O)$ and h_2 which possess very similar energies. The $2p_x$ and $2p_y$ orbitals of the oxygen atom remain as a doubly degenerate pair of π_u orbitals. The MO diagram for the linear water molecule is shown in Figure 5.11.

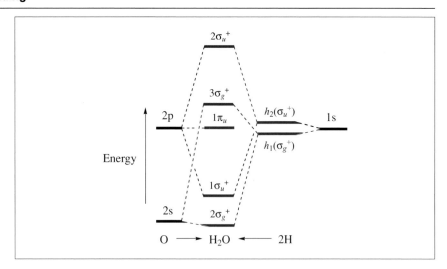

5.11 MO diagram for the linear water molecule

Comparison and Correlation of the Energies of 90° and 180° Water Molecules

In the water molecule there are eight valency electrons to be distributed in the lowest four MOs. In the 90° case this produces an electronic configuration which may be written as $2a_1{}^2 1b_2{}^2 3a_1{}^2 1b_1{}^2$ (the $1s^2$ pair of the oxygen atom forms the configuration $1a_1{}^2$ or K). The $2a_1$ orbital is slightly bonding because of the interaction with the $3a_1$ orbital. The latter orbital has considerable bonding character, as has the $1b_2$, so that there are two bonding pairs of electrons responsible for the cohesion of the three atoms. The $1b_1$ orbital is non-bonding, there being no hydrogen orbitals of that symmetry. The 90° molecule would have two similar, strongly bonding, electron pairs (in the $3a_1$ and $1b_2$ three-centre MOs) and two virtually non-bonding electron pairs (in the $1b_1$ and $2a_1$ orbitals) of very different energies.

The electronic configuration of the linear water molecule is $(2\sigma_g{}^+)^2(1\sigma_u{}^+)^2(1\pi_u)^4$ [the $1s^2(O)$ pair of electrons occupy the $1\sigma_g{}^+$ molecular orbital]. The $2\sigma_g{}^+$ pair of electrons is only weakly bonding, and the $1\sigma_u{}^+$ pair supplies practically the only cohesion for the three atoms, the other four electrons being non-bonding.

A comparison of the MO diagrams for the two forms of the water molecule is given in the **correlation diagram** of Figure 5.12, which correlates the MOs of the two extreme geometries of the molecule. It shows that the 90° angle confers the extra stability of two bonding pairs of electrons as opposed to the single pair of bonding electrons in the linear molecule.

The diagrams of Figures 5.10 and 5.11 have been constructed according to the convention of making the major axis coincident with the z axis. In considering the correlation of the two diagrams, it is useful to

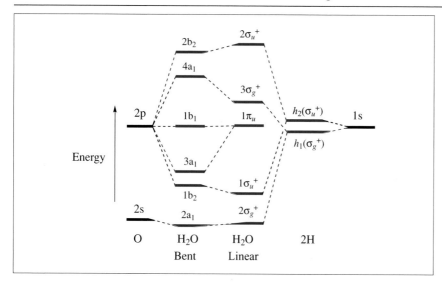

Figure 5.12 A MO correlation diagram for the bent and linear forms of the water molecule

retain the convention for the linear molecule, but to have the linear molecule bend in the yz plane so that the C_2 axis of the bent molecule is coincident with the y axis. In this case, the $1\pi_u$ orbitals of the linear molecule (oxygen atom contributions from $2p_x$ and $2p_y$ atomic orbitals) correlate with the $1b_1$ and $3a_1$, respectively, of the bent molecule. The $2p_y$ and $2p_z$ orbitals in the bent molecule are labelled a_1 and b_2, rather than the other way round which is conventional. The $2p_z$ orbital correlates with the $1b_2$ orbital of the bent molecule; it takes the place of the $2p_y$ orbital in the conventional treatment. These non-conventional changes may be visualized by swapping the y and z axes in Figures 5.8 and 5.9. The correlations are: $2\sigma_u^+ \rightarrow 2b_2$, $3\sigma_g^+ \rightarrow 4a_1$, $1\pi_u \rightarrow 1b_1$ and $3a_1$, $1\sigma_u^+ \rightarrow 1b_2$, and $2\sigma_g^+ \rightarrow 2a_1$.

Neither interelectronic repulsions nor internuclear repulsions have been considered. To ignore interelectronic repulsions is not serious since the orbitals used in the two forms of the molecule are extremely similar. The internuclear repulsion in the 90° form would be larger than in the linear case, and contributes to the bond angle in the actual water molecule being greater than 90°. The actual state of the molecule, as it normally exists, is that with the lowest total energy and only detailed calculations can reveal the various contributions. At a qualitative level, as carried out so far in this section, the decision from MO theory is that the water molecule should be bent, in preference to being linear.

The $1\pi_u$ orbitals in the linear case (these are the non-bonding $2p_x$ and $2p_y$ orbitals of the oxygen atom) *lose their degeneracy when the molecule bends*. The $2p_x$ orbital retains its non-bonding character, but the $2p_y$ orbital makes a very important contribution to the $3a_1$ bonding MO of the bent molecule. It is this factor which is critical in determining the shape of the water molecule. The distortion of the linear molecule produces two orbitals, labelled $4a_1$ and $3a_1$ in Figure 5.12, which are derived from the $3\sigma_g^+$ orbital and a component of the $1\pi_u$ orbital. The two orbitals in the bent molecule, having the same symmetry, can mix efficiently as they were quite close in energy in the linear molecule, and produce a stabilization of the lower orbital, $3a_1$, which then becomes a bonding orbital.

Experimental confirmation of the order of MO energies for the water molecule is given by its photoelectron spectrum. Figure 5.13 shows the helium-line photoelectron spectrum of the water molecule. There are three ionizations at 1216, 1322 and 1660 kJ mol^{-1}. A fourth ionization at 3107 kJ mol^{-1} has been measured by using suitable X-ray photons instead of the helium emission. That there are the four ionization energies is consistent with expectations from the MO levels for a bent C_{2v} molecule (see Figure 5.12).

In a band of peaks in a photoelectron spectrum, the peaks which represent vibrational excitations of the molecular ion in addition to the ionization of the molecule may be more intense than that which produces the molecular ion in its lowest energy vibrational state. It is the latter signal that corresponds to the ionization energy of the molecule.

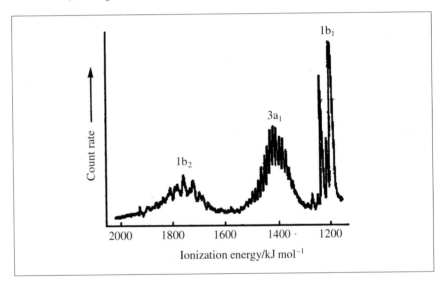

Figure 5.13 The photoelectron spectrum of water vapour; ionizations from the $1b_1$, $3a_1$ and $1b_2$ orbitals are indicated. Ionizations from the more stable $2a_1$ orbital are not produced by the helium radiation used

A study of the magnitudes of the various vibrational excitations associated with the second and third ionizations confirm that the second ionization is from the $3a_1$ orbital (this is both O–H and H–H bonding) and that the third ionization is from the $1b_2$ orbital (this is O–H bonding but H–H anti-bonding).

Worked Problem

Q If an electron is removed from the $1b_1$ orbital of the water molecule, what effect would this have on the bond angle? What would be the VSEPR prediction for the same ionization?

A The correlation diagram of Figure 5.12 shows that the $1b_1$ orbital of the water molecule is not dependent upon bond angle, so the removal of an electron from that orbital should not alter the bond angle in the resulting H_2O^+ ion. This is in accord with experiment. The VSEPR theory would predict that the loss of an electron from one of the two non-bonding orbitals would cause the H–O–H angle to open up as there would be less repulsion from the position containing the single electron.

Conclusion

The discussion of the distortion of the water molecule from a linear to a bent shape allows a *tentative general conclusion* to be reached. This is that if a distortion of a molecule from a particular symmetry allows two MOs to mix, so that the lower occupied orbital is stabilized at the expense of the higher vacant orbital, such a distortion will occur and will confer stability on the distorted molecule. A gain of stability will only occur if the two orbitals concerned in the stabilization process are the highest occupied molecular orbital (HOMO) and the lowest unoccupied molecular orbital (LUMO). If both orbitals are doubly occupied, interaction between them does not lead to any change in stability. The generality of this conclusion is explored further in the next sections of this chapter and in Chapter 6.

5.3.2 Molecular Orbital Theory of CO_2 and XeF_2

The linear CO_2 and XeF_2 molecules belong to the $D_{\infty h}$ point group and, for purposes of classifying their atomic orbitals, the principal quantum numbers (2 for C, 5 for Xe) of their *valence electrons* are omitted in the following discussion.

The s orbitals of C and Xe transform as σ_g^+, the p_x and p_y transform as π_u, and the p_z as σ_u^+. The p–s energy gaps in oxygen and fluorine atoms are very large and their 2s orbitals (as formal bonding and anti-bonding combination group orbitals) may be regarded as being virtually non-bonding. Using the Mulliken numbering system for the CO_2 case, the 1s orbitals of the ligand atoms form the $1\sigma_g^+$ and $1\sigma_u^+$ combinations. The 1s orbital of the central carbon atom is next highest in energy and is labelled as $2\sigma_g^+$. The 2s orbitals of the ligand atoms give rise to the combinations labelled as $3\sigma_g^+$ and $2\sigma_u^+$.

The 2p orbitals of the ligand atoms, labelled as either A or B to distinguish between the two atoms, may be dealt with in two sets of group orbitals: the σ orbitals being the $2p_z$ lying along the molecular axis, and the π orbitals being made up from the $2p_x$ and $2p_y$ orbitals perpendicular to the molecular axis. The 2p orbital group combinations may be written as:

$$\phi(\sigma_g^+) = \psi(2p_z)_A + \psi(2p_z)_B \qquad (5.1)$$

the *plus sign* being used to indicate a *bonding interaction* (*i.e.* a plus ψ to plus ψ overlap), and

$$\phi(\sigma_u^+) = \psi(2p_z)_A - \psi(2p_z)_B \qquad (5.2)$$

the *minus sign* indicating an *anti-bonding interaction* (*i.e.* a plus ψ to minus ψ overlap).

The π group orbital combinations are:

$$\phi(\pi_u) = \psi(2p_{x,y})_A + \psi(2p_{x,y})_B \qquad (5.3)$$

and

$$\phi(\pi_g) = \psi(2p_{x,y})_A - \psi(2p_{x,y})_B \qquad (5.4)$$

with the signs having the same significance as those in equations 5.1 and 5.2. A summary of the above classifications is given in Table 5.3.

Table 5.3 The orbitals of C and Xe and the group orbitals of O and F which can combine to give the molecular orbitals of CO_2 and XeF_2. The references to bonding characteristics in the table are with respect to interactions between the two ligand atoms

C or Xe atomic orbital	O_2 or F_2 group orbitals
	σ_g^+ (2s ~non-bonding)
	σ_u^+ (2s ~non-bonding)
σ_g^+ (s)	σ_g^+ ($2p_z$ bonding)
σ_u^+ (p_z)	σ_u^+ ($2p_z$ anti-bonding)
π_u ($p_{x,y}$)	π_u ($2p_{x,y}$ bonding)
	π_g ($2p_{x,y}$ anti-bonding)

It is possible to use one qualitative MO energy diagram for CO_2 and XeF_2 molecules. The diagram is shown in Figure 5.14 and for CO_2 omits the $1\sigma_g^+$, $1\sigma_u^+$, $2\sigma_g^+$, $3\sigma_g^+$ and $2\sigma_u^+$ orbitals (*i.e.* the two ligand 1s orbitals, the 1s orbital of the central carbon atom and the two ligand 2s orbitals). The orbital numbering by the Mulliken system takes all the orbitals into account; hence the numbering on the diagram of Figure 5.14 has the bonding combination of the carbon 2s $[\sigma_g^+(s)]$ orbital with the O–O bonding $2p_z$ oxygen group orbital $[\sigma_g^+(2p_z \text{ bonding})]$ as the $4\sigma_g^+$ MO. The π MOs are assumed to be between the bonding and anti-bonding σ orbitals, as the π overlap is normally not as efficient as that of the σ atomic orbitals.

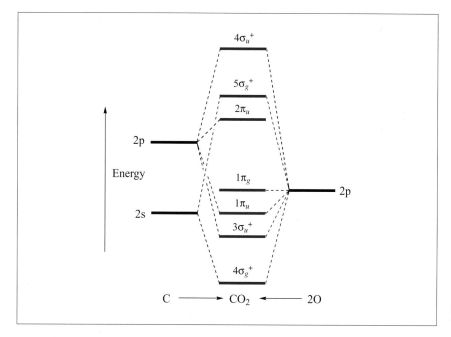

Figure 5.14 A MO diagram for the CO_2 and XeF_2 molecules

The 16 valence electrons of the CO_2 molecule (*i.e.* $2s^22p^2$ from C and $2s^22p^4$ from both Os) occupy the lowest energy orbitals, in accord with the *aufbau* principle, to give the electronic configuration as $(3\sigma_g^+)^2(2\sigma_u^+)^2(4\sigma_g^+)^2(3\sigma_u^+)^2(1\pi_u)^4(1\pi_g)^4$. The first two pairs of electrons are non-bonding (they are the $2s^2$ pairs of the oxygen atoms), but the next four pairs (two σ and two π) occupy three-centre (*i.e.* all three atoms participating) bonding MOs and are responsible for the high bond strength of the so-called 'double bonds' of the molecule (743 kJ mol^{-1} for each nominal C=O bond). The remainder of the electrons occupy the three-centre non-bonding π_g orbitals. The photoelectron spectrum of the CO_2 molecule indicates ionization energies of 1330, 1671, 1744 and 1872 kJ mol^{-1}, for the removal of electrons from the $1\pi_g$, $1\pi_u$, $3\sigma_u^+$ and $4\sigma_g^+$ MOs, respectively.

Some comment is necessary here to emphasize the three-centre nature of the bonding in CO_2, particularly with respect to the π orbitals. The three p orbitals in the xz plane and the three in the yz plane (remember that the molecular axis is conventionally taken to coincide with the z axis) form the three-centre set of three π MOs, $1\pi_u$ (bonding), $1\pi_g$ (non-bonding) and $2\pi_u$ (anti-bonding), the first two orbitals being fully occupied and leading to the idea that the three atoms are bonded by two three-centre π bonds. The valence bond theory, which depends upon two-electron two-centre bonding, provides the CO_2 molecule with two C=O double bonds (with the two π C–O bonds at right angles to each other with reference to the molecular axis) and can cope with the *delocalization of the π system* only by postulating the participation of ionic structures, $^-$O–C≡O$^+$ and $^+$O≡C–O$^-$, in the resonance hybrid. This ensures, in a clumsier manner than does MO theory, that the doubly degenerate π bonding is delocalized over the three atom centres. The delocalization allows the electrons of the π system to occupy larger orbitals than they would occupy in a two-electron, two-centre conventional bond. This allows a reduction in the interelectronic repulsion and leads to stronger bonding and stabilizes the molecule as a whole. This point can be demonstrated by a simple calculation. The standard enthalpy of formation of the CO_2 molecule is –394 kJ mol^{-1}. An estimate of what this would be if the molecule were to be constructed from two-electron, two-centre bonds (two σ and two π bonds) is calculated by adding the enthalpy changes for (i) the sublimation of carbon (715 kJ mol^{-1}), (ii) the dissociation of dioxygen (496 kJ mol^{-1}) and (iii) the production of two C=O bonds from one carbon atom and two oxygen atoms (2 × –743 kJ mol^{-1}). The calculated enthalpy of formation of the CO_2 molecule is thus –275 kJ mol^{-1}, which indicates that the localized two double bonded molecule would be 119 kJ mol^{-1} less stable than the real delocalized arrangement. The energy changes described are shown in Figure 5.15.

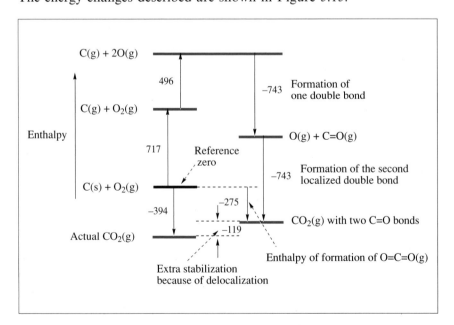

Figure 5.15 The energy changes in the formation of carbon dioxide from its elements

---

Box 5.4 Delocalization

Delocalization occurs in molecules wherever it is possible by symmetry considerations and wherever an energetic advantage can be gained from its operation. MO theory deals very satisfactorily with delocalization, but with valence bond theory the concept is somewhat clumsily incorporated as an addition to the conventional two-electron, two-centre bonding. In general, the stability conferred upon a molecule by delocalization is because the orbitals are more extensive, so that interelectronic repulsion is minimized.

The electronic configuration of the 22-electron XeF_2 molecule may be deduced from the diagram of Figure 5.14 and may be written (using the same prefixes as those for CO_2) as $(4\sigma_g^+)^2(3\sigma_u^+)^2(1\pi_u)^4(1\pi_g)^4(2\pi_u)^4(5\sigma_g^+)^2$ for the 18 valency electrons (*i.e.* ignoring the two 2s pairs of electrons of the fluorine atoms). It follows the same pattern as that of the CO_2 molecule with the additional occupancy of the $3\sigma_g^+$ and the doubly degenerate $2\pi_u$ orbitals, all three orbitals being anti-bonding. Their occupancy reduces the resultant bonding in the XeF_2 molecule to the single electron pair in the $3\sigma_u^+$ bonding orbital. The bonding of the molecule is equivalent to each Xe–F 'bond' possessing a bond order of 0.5. The molecule is only marginally stable with respect to its formation, $\Delta_f H(XeF_2,g) = -82$ kJ mol^{-1}, compared with CO_2, $\Delta_f H^\ominus(CO_2,g) = -394$ kJ mol^{-1}. A fairer comparison would be to use the standard heat of formation of CO_2 from carbon in its gaseous state which is -1109 kJ mol^{-1}. The XeF_2 molecule would be difficult to describe in terms of valence bond theory since conventional electron pair bonds are impossible with the filled shell Xe atom, and in MO theory the effects of π-electron delocalization are not present because of the filled $2\pi_u$ anti-bonding orbitals.

The He-line photoelectron spectrum of XeF_2 shows five ionization energies at 1206, 1322, 1389, 1515 and 1679 kJ mol^{-1}, which have been assigned (by a study of the vibrational intervals) to the removal of electrons from the $2\pi_u$, $5\sigma_g^+$, $1\pi_g$, $1\pi_u$ and $3\sigma_u^+$ orbitals, respectively. The reversal of the energies of the $2\pi_u$ and $5\sigma_g^+$ orbitals, compared to those indicated in Figure 5.14, is possibly because of d-orbital participation in XeF_2, which is not possible in CO_2. The MO treatment of XeF_2 does not depend upon (and so does not overemphasize) the inclusion of d-orbital contributions from the Xe atom. Reference to the $D_{\infty h}$ character table shows that the d_{z^2} orbital of xenon transforms as σ_g^+ and that the d_{xz} and d_{yz} orbitals transform as π_g. Those orbitals have suitable symmetries to interact with the appropriate orbitals of the fluorine atoms, so that the orbitals labelled $5\sigma_g^+$ and $1\pi_g$ could derive some stabilization by 5d participation. The other two d orbitals (xy and x^2–y^2) transform as the representation, δ_g, and cannot participate in the bonding of XeF_2.

An alternative and simpler view of the bonding in XeF_2 is that using three-centre, four-electron bonds (3c4e) consisting of orbitals constructed from the three p_z atomic orbitals of the three atoms. A diagram of their overlap is shown in Figure 5.16. The $5p_z$ orbital of the xenon atom transforms as σ_g^+, as does the anti-bonding combination of the two $2p_z$ orbitals of the ligand fluorine atoms. The interaction of these two orbitals gives rise to the bonding and anti-bonding combinations shown in Figure 5.17. The non-bonding MO is that produced by the bonding combination of the two fluorine orbitals with σ_u^+ symmetry. The xenon $5p_z$ orbital contains two electrons and the fluorine $2p_z$ orbitals are singly occupied. The bonding is produced by the four electrons occupying the σ_g^+ bonding MO and the non-bonding σ_u^+ MO, leading again to a Xe–F bond order of 0.5. This σ-only treatment of the bonding in XeF_2 leads to the same conclusions as the more complex MO treatment, and underlines the point that it is unnecessary to have any participation from the 5d level.

This 3c4e approach may be used to understand the bonding and shape of xenon tetrafluoride, XeF_4, which is square planar (D_{4h}), but it is inadequate for XeF_6 which is not a regular octahedron, and shows evidence of the effects of the *seven* electron pairs which the valence shell of the xenon atom contains.

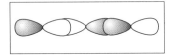

Figure 5.16 Overlap diagrams for the p orbitals along the z axis of the XeF_2 molecule

Figure 5.17 The three-centre orbitals of the XeF_2 molecule

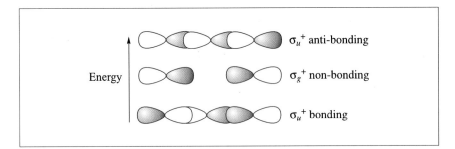

σ_u^+ anti-bonding

Energy

σ_g^+ non-bonding

σ_u^+ bonding

Worked Problem

Q Use Figure 5.14 to determine the electronic configuration of the BrF_2^- ion, and compare your results with the text concerning the XeF_2 molecule.

A Figure 5.14 shows the MOs produced from the valency electrons of the central ion plus the valence 2p electrons of the ligand fluorine atoms. For the BrF_2^- ion there would be the seven electrons from the bromine atom, the six 2p electrons from the two fluorine atoms and the extra single electron responsible for the negative charge, making a total of 18 electrons to be disposed of in the lowest nine levels of the figure. This is an identical situation to that described for XeF_2 and it would be expected that the BrF_2^- ion would be linear.

5.3.3 Molecular Orbital Theory of NO_2^+, NO_2 and NO_2^-

The nitronium cation, NO_2^+, is linear, but the +4 oxide of nitrogen, NO_2, and the nitrite [nitrate(III)] ion, NO_2^-, are both bent structures and belong to the C_{2v} point group. The bond angles and bond lengths for NO_2^+, NO_2 and NO_2^- are given in Table 5.4. The three species form an important application of MO theory as it deals with bond angles in cases where there is π bonding.

Table 5.4 Data for NO_2^+, NO_2 and NO_2^-

Species	N—O length/pm	O—N—O angle
NO_2^+	115	180°
NO_2	119	134°
NO_2^-	124	115°

Unlike the molecules in the previous section, the 2p–2s energy gap in the central atom is relatively large, and this has an effect on the participation of the 2s orbitals of the nitrogen and oxygen atoms in MO formation. The MO diagram for NO_2^+, shown in Figure 5.18, differs somewhat from that of Figure 5.14 for CO_2 because the 2p–2s energy gap is much larger in the nitrogen atom. The MOs may be visualized by taking the orbital diagrams of Figure 4.1 for E_2 molecules, stretching

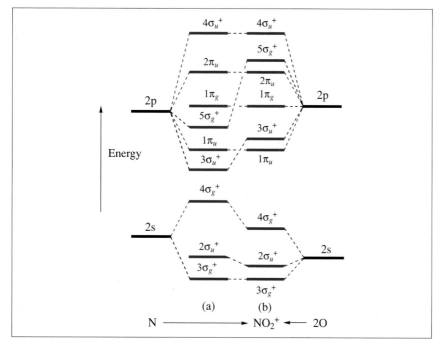

Figure 5.18 The MO diagram for NO_2^+

them to allow for an atom to be placed into their centres, and then slotting in the 2s or 2p orbitals of the central atom, as appropriate.

Because the first ionization energy of the oxygen atom (1310 kJ mol^{-1}) is slightly lower than that of the nitrogen atom (1400 kJ mol^{-1}), and because the 2p–2s energy gap in O (1500 kJ mol^{-1}) is somewhat larger than that in N (1050 kJ mol^{-1}), the energies of the 2s atomic orbitals of the nitrogen and two oxygen atoms are not very different and MO interaction is possible. The MO diagram may be described in two parts, the first assuming no MO mixing (of those orbitals with identical symmetry), and then in the second part such orbital mixing is incorporated.

Working up the diagram in Figure 5.18(a) from the most stable MO, the nitrogen 2s orbital interacts with the σ_g^+ group combination of the oxygen 2s orbitals to give the bonding and anti-bonding MOs labelled $3\sigma_g^+$ and $4\sigma_g^+$, respectively. The anti-bonding combination of the oxygen 2s orbitals is N–O non-bonding and is labelled $2\sigma_u^+$. The σ interactions between the $2p_z$ orbital of the nitrogen atom and the σ_u^+ combination of the oxygen $2p_z$ orbitals give the bonding and anti-bonding MOs labelled $3\sigma_u^+$ and $4\sigma_u^+$ in Figure 5.18. The remaining σ MO is that which is labelled $5\sigma_g^+$ and consists of the bonding combination of the oxygen $2p_z$ atomic orbitals, which does not interact with any orbitals of the nitrogen atom at this first stage. Next come the doubly degenerate interactions of the $2p_x$ and $2p_y$ orbitals of the nitrogen atom with the π_u combinations of the corresponding orbitals of the oxygen atoms to give the bonding and anti-bonding MOs labelled $1\pi_u$ and $2\pi_u$. The remaining π orbitals are the doubly degenerate π_g combinations of the $2p_x$ and $2p_y$ orbitals of the oxygen atoms, which are non-bonding and have no interaction with nitrogen atomic orbitals and are labelled $1\pi_g$ in Figure 5.18.

When orbital mixing is considered, it affects the energies of the orbitals with identical symmetry and produces the changes shown in Figure 5.18(b). The $4\sigma_g^+$ and $5\sigma_g^+$ orbitals mix to stabilize the lower energy orbital at the expense of the higher one. This causes the $5\sigma_g^+$ orbital to be higher in energy than the $2\pi_g$ set. The anti-bonding $4\sigma_g^+$ orbital is correspondingly stabilized and becomes less anti-bonding. In a similar manner, the interaction between the $2\sigma_u^+$ and $3\sigma_u^+$ orbitals causes the bonding $3\sigma_u^+$ orbital to become destabilized as the lower energy $2\sigma_u^+$ orbital gains stability. The $3\sigma_u^+$ loses some of its bonding character and rises above the $1\pi_u$ level. The crucially important interaction is that between the $4\sigma_g^+$ and $5\sigma_g^+$ orbitals, which prevents the latter orbital being occupied in the nitronium ion. Hybridization allows the stabilization of the species in this case. That between the $2\sigma_u^+$ and $3\sigma_u^+$ orbitals is relatively unimportant because both orbitals are occupied and no change of energy results from their mixing. The mixing of the $4\sigma_g^+$ and $5\sigma_g^+$ orbitals is shown pictorially in Figure 5.19.

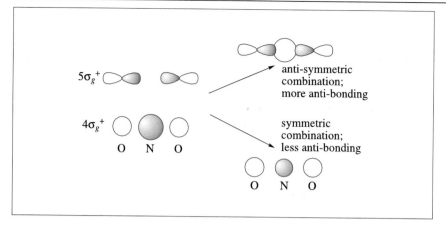

Figure 5.19 The interactions between the $4\sigma_g^+$ and $5\sigma_g^+$ orbitals of NO_2^+. The interactions between the oxygen orbital contributions are minimal, and are not shown

Box 5.5 Drawbacks of Single Diagrams

It should be realized that the attempt to derive one diagram which will suffice for all examples of triatomic molecules is almost impossible without having to make particular changes for some molecules. This is because of the varying electronegativities of the central and ligand atoms and of their relative s–p energy gaps. The varying first ionization energies which mirror the electronegativity coefficients of the atoms decide the availability of the sets of s and p orbitals to combine with each other, the most compatible situation being that where the atomic orbital energies are as close as possible. The relative magnitudes of the s–p energy gaps govern the extent of orbital mixing that may occur between MOs of identical symmetry.

In NO_2^+ there are 16 valency electrons which occupy the lower six levels in Figure 5.18(b) to give the electronic configuration $(3\sigma_g^+)^2(2\sigma_u^+)^2(4\sigma_g^+)^2$ $(1\pi_u)^4(3\sigma_u^+)^2(1\pi_g)^4$, equivalent to a formal bond order of 3.0 arising from the three pairs of bonding electrons in the $3\sigma_u^+$ and $1\pi_u$ orbitals. Because of the stabilization of the $4\sigma_g^+$ orbital it may be regarded as almost non-bonding, so that the bonding effect of the electron pair in the $3\sigma_g^+$ orbital gives a boost to the bond order by another unit. This means that the two bonds in the nitronium ion may be considered to be double bonds, but it is important to realize that the four pairs of bonding electrons are delocalized over the three atom centres to give an extra stability to the ion.

NO_2 and NO_2^-

If NO_2 and NO_2^- were linear, their valency electrons in excess of 16 would occupy the anti-bonding $2\pi_u$ anti-bonding level, $(2\pi_u)^1$ for NO_2 and $(2\pi_u)^2$ for NO_2^-. By having bent structures they avoid the occupancy of anti-bonding orbitals, and thereby achieve a greater stability than they would have if they were linear. In constructing the MO diagram shown in Figure 5.20 it is important to note how the MOs of the linear molecules shown in Figure 5.18(b) transform with respect to the new symmetry of the C_{2v} point group to which the bent molecules belong.

Starting with the lowest energy orbitals and working upwards, in Figure 5.20 the $3\sigma_g^+$ orbital becomes the $3a_1$ orbital. Bending brings the ligand atoms together, and in this case the resulting orbital becomes more stable since the contributions from the oxygen 2s orbitals are in a bonding mode. The $2\sigma_u^+$ orbital becomes the $2b_2$ of the bent molecules and is slightly higher in energy because of the anti-bonding mode of the oxygen 2s orbitals from which this orbital is constructed. The $4\sigma_g^+$ orbital becomes the $4a_1$ orbital of the bent molecules, deriving a little stability because the oxygen 2s contributions are in a bonding mode. The doubly degenerate $1\pi_u$ MOs of the linear molecules lose their degeneracy in the bent cases, with transformation into the orbitals labelled $1b_1$ and $5a_1$, the former orbital remaining as a π-type orbital. The $5a_1$ orbital is at a higher level then the $1\pi_u$, as it has lost some of its bonding character.

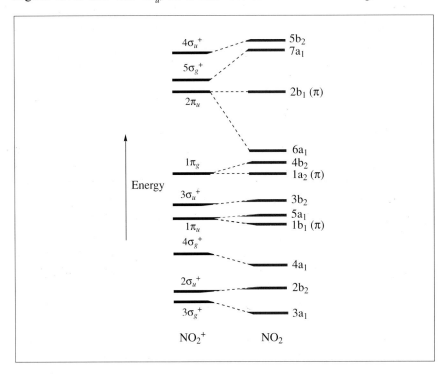

Figure 5.20 A MO diagram for bent NO_2 species

The $3\sigma_u^+$ orbital becomes the $3b_2$ of the bent molecules and increases in energy slightly because it becomes less N–O bonding, and the anti-bonding mode of the oxygen 2p orbitals from which this orbital is constructed approach more closely and are more anti-bonding. The non-bonding $1\pi_g$ set of degenerate orbitals splits upon bending into the $4b_2$ and $1a_2$ pair of orbitals, the former having a slightly higher energy because the oxygen 2p contributions are in an anti-bonding mode. The $1a_2$ orbital is also destabilized since it is more O–O anti-bonding in the bent cases.

So far, the changes described apply to orbitals which are all occupied in the molecules under discussion, and therefore do not alter the overall stability of the molecules substantially. This is not the case for the anti-bonding $2\pi_u$ orbitals, which in NO_2 and NO_2^- represent the HOMOs. They lose their degeneracy by bending, to give the orbitals labelled as $2b_1$ and $6a_1$. The former remains as an anti-bonding π-type orbital, and the latter is approximately non-bonding and *initially* derives some stabilization as a result. The $5\sigma_g^+$ orbital is the LUMO for NO_2 and NO_2^-, and which in the linear molecule is a bonding combination of the two oxygen 2p orbitals, is *initially* regarded to be unchanged in energy in the bent examples where it becomes labelled as $7a_1$. The highest energy orbital in the linear case is $4\sigma_u^+$ and becomes the $5b_2$ orbital of the bent molecules. The oxygen 2p contributions are in an anti-bonding mode, but there is a loss of some N–O anti-bonding character so the orbital is slightly destabilized in the bent molecules.

Having dealt with the transformation of all the MOs of the linear molecules to those of the bent variety, there is a very important *extra effect* which is incorporated into Figure 5.20 as the result of the interaction of the orbitals $6a_1$ and $7a_1$. Because these orbitals have the *same symmetry* and are fairly close to each other in energy, they may and do interact so that the $6a_1$ orbital is stabilized at the expense of the $7a_1$ orbital. The atomic orbital contributions to the MOs are shown in Figure 5.21 for the $2\pi_u$–$6a_1$ and $5\sigma_g^+$–$7a_1$ transformations.

The interaction between the $3b_2$ and $4b_2$ orbitals stabilizes the former at the expense of the latter, but since the two orbitals are both occupied in the molecules being discussed there is no change in their total energies and the interaction is not important. The $6a_1$–$7a_1$ interaction has consequences for the electronic configurations of NO_2 and NO_2^- and their bond angles. The highest energy electrons in NO_2 and NO_2^- have the configurations $(6a_1)^1$ and $(6a_1)^2$, respectively, and since this orbital is strongly stabilized upon bending, the bond angles are consistent with the transition from the linear NO_2^+ to the bent NO_2 and NO_2^-, the latter having a smaller bond angle than the former since it has two electrons in the $6a_1$ orbital.

As in the case of the distortion of the linear water molecule (Section 5.2.1), this more complicated treatment of NO_2 species arrives at the con-

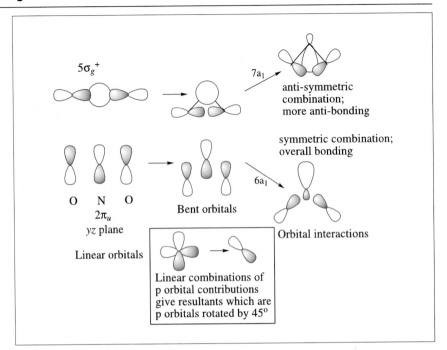

Figure 5.21 The $2\pi_u$–$6a_1$ and $5\sigma_g^+$–$7a_1$ orbital transformations which occur when NO_2^+ bends

clusion that the distortion (*i.e.* bending) of the NO_2 and NO_2^- species occurs because of the stabilization of their HOMOs. The stabilization is possible because the distortion allows the HOMO and LUMO of the bent molecules to have the same symmetry, and therefore they mix, giving stability to the lower of the two orbitals and to the molecule, since the higher orbital is vacant.

5.3.4 XeF$_2$ again

In Section 5.2.2 the electronic configuration of the XeF_2 molecule is treated as being very similar to that of the linear molecule CO_2. That is because they are both linear molecules. It is clear that CO_2 should be linear from the electronic configuration derived in Section 5.2.2, in which the C–O bond orders are shown to be 2 and the inter-ligand repulsion is minimized. In XeF_2, however, the bond orders are a mere 0.5, and some reason should be furnished for its linearity. Electrons occupy antibonding orbitals, $(2\pi_u)^4(5\sigma_g^+)^2$, and surely some distortion should occur which gives rise to some stabilization, as it does in the cases of NO_2 and NO_2^-. There is no alleviation of instability for XeF_2 even if it is distorted to a bent state, because the HOMO would then be the doubly occupied $7a_1$ orbital, which would cancel out the stabilization of having the $6a_1$ orbital occupied. The molecule is linear because of the minimization of inter-ligand repulsion, rather than any electronic cause.

5.3.5 Conclusions

Predictions of the shapes of triatomic molecules, based on the above ideas, indicate that 16-electron molecules should be linear (no anti-bonding orbitals occupied), but molecules with 17–20 electrons will be bent because of electronic stabilization. Molecules with 21 electrons should have the larger angle associated with the 17-electron molecules, the latter having the configuration $(6a_1)^1$ and the former having $(6a_1)^2(2b_1)^2(7a_1)^1$. Molecules with 22 electrons will be linear because of the absence of electronic stabilization and the minimization of inter-ligand repulsion. Table 5.5 gives examples of molecules having numbers of valence electrons from 16 to 22 and their electronic configurations and bond angles.

Table 5.5 Bond angles and electronic configurations for molecules with 16–22 electrons

Example	Number of valence electrons	$6a_1$	$2b_1$	$7a_1$	Bond angle
CO_2	16	0	0	0	180°
NO_2	17	1	0	0	134°
$ONCl$	18	2	0	0	116°
ClO_2	19	2	1	0	117°
OCl_2	20	2	2	0	111°
XeF_2^+	21	2	2	1	134°
XeF_2	22	2	2	2	180°

The examples given in Table 5.5 are molecules whose electronic configurations are reasonably deduced from Figures 5.14, 5.18 and 5.20. Table 5.6 gives bond length and bond angle data for some compounds with the formula EL_2, where E can be O, S or Se and L can be H, F or Cl.

Because the relevant s and p orbitals of sulfur and selenium are closer together and at higher general energies than the 2s and 2p orbitals of oxygen, it is not possible to use any of the diagrams of this chapter to discuss their electronic configurations. Some general comments may be made about the trends in bond angles. In the series of OL_2 compounds there is very little correlation of bond angle with the size of the ligand atoms, although there is some correlation in the quantities for the SL_2 compounds. Across the series of EH_2, EF_2 and ECl_2 compounds there is a reduction in the bond angle as the size of the central atom increases. The limit of the size reduction seems to be 90°, corresponding to the full participation of the two p orbitals in the bonding and to the

Table 5.6 Bond length and bond angle data for some compounds with the formula EL_2. The differences between the electronegativity coefficients (Allred–Rochow) of the elements are also given

Property	OH_2	SH_2	SeH_2
Angle	104.5°	92°	91°
r(F–H)/pm	96	133	146
r(H–H)/pm	152	192	208
$\delta\chi$	1.3	0.2	0.3

	OF_2	SF_2
Angle	103.8°	98°
r(E–F)/pm	141	159
r(F–F)/pm	222	240
$\delta\chi$	0.6	1.7

	OCl_2	SCl_2
Angle	110.8°	103°
r(E–Cl)/pm	170	200
r(Cl–Cl)/pm	282	313
$\delta\chi$	0.7	0.4

maximization of electronic stabilization. This latter factor overrides the effects of inter-ligand repulsion. There is no noticeable correlation between the physical data and the differences in the electronegativity coefficients of the participating elements in the compounds considered.

From a VSEPR viewpoint it might be expected that the non-bonding pairs would have less effect on reducing the bond angles from the expected regular tetrahedral angle as the size of the central atom increased. It might also be expected that as the ligand atoms become more electronegative they would have a greater share of the bonding electrons and would repel each other more strongly, leading to a larger bond angle. Both predictions from VSEPR theory seem not to hold.

5.4 Hydrogen Bonding and the Hydrogen Difluoride ion, HF_2^-

The hydrogen difluoride ion has a central hydrogen atom and belongs to the $D_{\infty h}$ point group. It is important in explaining the weakness of hydrogen fluoride as an acid in aqueous solution, and as an example of **hydrogen bonding**. Hydrogen bonding occurs usually between molecules which contain a suitable hydrogen atom and one or more of the very electronegative atoms N, O or F. It is to be considered, in general, as a particularly strong intermolecular force.

In the case of HF_2^- it may be considered that the ion is a combination of an HF molecule and a fluoride ion, F^-. Both of those chemical entities have considerable thermodynamic stability independently of each other. Their combination in HF_2^- indicates the operation of a force of attraction which is significantly greater than the normal intermolecular forces which operate in all systems of molecules. The latter forces are responsible for the stability of the liquid and solid phases of chemical substances. They counteract the tendency towards chaos. The majority of hydrogen bonding occurs in systems where the hydrogen atom, responsible for the bonding, is asymmetrically placed between the two electronegative atoms (one from each molecule). In those cases the bonding may be regarded as an extraordinary interaction between two dipolar molecules. The extraordinary nature of the effect originates in the high effectiveness of the almost unshielded hydrogen nucleus in attracting electrons from where they are concentrated: around the very electronegative atoms of neighbouring molecules. The hydrogen bond in the HF_2^- ion is capable of treatment by bonding theories.

The VSEPR treatment is best approached by considering the ion as made up from three ions: $F^- + H^+ + F^-$. The central proton possesses no electrons until the ligand fluoride ions supply two each. The two pairs of electrons repel each other to give the observed linear configuration of the three atoms. The two pairs of electrons would occupy the 1s and 2s orbitals of the hydrogen atom and, what with a considerable amount of interelectronic repulsion, would not lead to stability.

The MO theory is straightforward. The hydrogen 1s orbital transforms as σ_g^+ within the $D_{\infty h}$ point group. The $2p_z$ orbitals of the fluorine atoms (since the ion is set-up with the molecular axis coincident with the Cartesian z axis) may be arranged in the linear combinations:

and

$$\phi(\sigma_g^+) = \psi(2p_z)_A + \psi(2p_z)_B \tag{5.5}$$

$$\phi(\sigma_u^+) = \psi(2p_z)_A - \psi(2p_z)_B \tag{5.6}$$

the As and Bs indicating that the 2p contributions are from the two fluorine atoms.

MOs may be constructed from the two orbitals of σ_g^+ symmetry to give bonding and anti-bonding combinations. The orbital, localized on the fluorine atoms, of σ_u^+ symmetry remains as a non-bonding orbital. The MO diagram is shown in Figure 5.22. The four electrons, which may be regarded to have been supplied by two fluoride ions, occupy the bonding and non-bonding orbitals and the F–H bond order is 0.5. This is consistent with the observed weakness of the hydrogen bond of 126 kJ mol^{-1} (the $\Delta H^{\ominus}$ of the reaction of HF_2^- to give $H^+ + HF$), although that is very large relative to 'normal' hydrogen bond strengths of 10–30 kJ mol^{-1}.

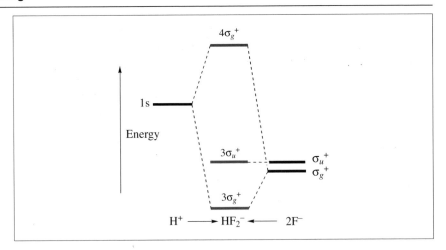

Figure 5.22 A MO diagram for the HF_2^- ion

Box 5.6 'Normal' Hydrogen Bonding

What has been referred to as 'normal' hydrogen bonding is not dealt with in this text, but some explanation is included here. Water melts at 0 °C and boils at 100 °C, both physical constants being abnormally high when compared to those of the dihydrides of the other elements of Group 16. Table 5.7 gives this information.

Table 5.7 Physical constants for some Group 16 dihydrides

Dihydride	Melting temperature/K	Boiling temperature/K
H_2S	–85	–61
H_2Se	–60	–41
H_2Te	–49	–2

Box 5.7 Van der Waals Forces

Van der Waals forces are intermolecular and are classified as (i) dipole–dipole interactions, (ii) dipole–induced-dipole interactions and (iii) London dispersion forces which operate between atoms as the result of the nucleus not always being at the centre of mass of the surrounding electrons. The hydrogen bond is regarded as a special form of dipole–dipole interaction, because the positive end of dipolar species containing hydrogen atoms is the relatively unshielded proton.

Over and above the normal van der Waals forces which are responsible for the cohesion of molecular compounds in the solid and liquid states, hydrogen bonding occurs between molecules containing electronegative atoms (*i.e.* mainly O and N) and hydrogen atoms. It may be regarded as an enhanced form of dipole–dipole interaction, as shown in Figure 5.23 for the water molecule, and in water is responsible for the arrangement of the molecules in the various forms of ice. When ice melts the liquid water retains some local order, which is finally broken down completely when the liquid boils.

The HF_2^- ion is isoelectronic with the unknown compound helium difluoride, HeF_2. The latter compound would have a very similar electronic configuration to that of HF_2^-. The reason for its non-existence is indicated by the values of the first ionization energies of H^- and He (which are isoelectronic) of 73 and 2372 kJ mol^{-1}, respectively. The attraction for electrons represented by the large ionization energy of the helium atom is greater than that exerted by the two fluorine atoms, making the formation of a stable compound impossible. The molecule XeF_2 exists (even though the first ionization energy of Xe is 1170 kJ mol^{-1}), although it is very weakly bonded.

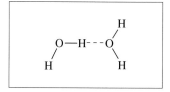

Figure 5.23 Dipole–dipole interaction between two water molecules as an example of hydrogen bonding

Summary of Key Points

1. Valence bond theory and molecular orbital theory were extended to cope with the bonding and bond angles of triatomic molecules.

2. Valence shell electron pair repulsion theory was applied to all the examples used in the chapter.

3. The detailed application of symmetry theory to the linear and bent forms of the water molecule was described.

4. The method of applying symmetry theory was discussed in terms of four stages of a general protocol for the derivation of the MO energy diagram.

5. MO theory of the linear species CO_2, XeF_2, I_3^-, NO_2^+, and the bent species NO_2 and NO_2^-, was discussed in detail.

6. The linear HF_2^- ion was treated as an example of strong hydrogen bonding.

7. The idea that distortion of a triatomic molecule from a linear to a bent shape occurs if the HOMO and LUMO are of the same symmetry representation, so that electrons in the HOMO are stabilized, was discussed as a tentative general approach to molecular shape.

Problems

5.1. With the help of electronic 'box' diagrams, show how VSEPR theory deals with the formation of NO_2^+, NO_2 and NO_2^-.

5.2. Apply VSEPR theory to the molecules BeF_2, CS_2 and the ion ClF_2^+ to give their shapes. Use Figures 5.14 and 5.20 to determine the shapes and bond orders of the bonds in the BeF_2 and CS_2 molecules and the ClF_2^+ ion.

5.3. The chlorine dioxide molecule, ClO_2, contains a chlorine atom in the unusual formal oxidation state (+4), and is an *odd* molecule containing 19 valency electrons. The molecule has a bond angle of 117° with a Cl–O bond length of 148 pm. Apply the VSEPR treatment to the molecule, and use Figure 5.20 to decide the MO electronic configuration. Compare the results from the two theories in their ability to explain the bond angle and the bond length. The single bond covalent radii of Cl and O are 99 and 74 pm, respectively.

Further Reading

J. K. Burdett, *Molecular Shapes*, Wiley, New York, 1980. This book covers all aspects of molecular shapes in a very readable and thorough manner.

6

Covalent Bonding IV: Polyatomic Molecules

Polyatomic molecules, in this book, are those with more than three atoms. The discussions in this chapter are restricted to a selection of small polyatomic molecules. Treatments of the shapes and bonding by VSEPR/valence bond theory and MO theories are included and compared. MO theory is presented in a less detailed fashion, and its main conclusions described rather than constructing the orbital diagrams, as is done in Chapter 3 for triatomic molecules. The molecules whose bonding schemes are described are (1) the two hydrides NH_3 and CH_4; (2) a series of tetra-atomic fluorides, BF_3, NF_3 and ClF_3 with varying numbers of valence electrons, treated in a less rigorous fashion and (3) some 'two-centre' molecules are described: B_2H_6 and ethane are compared. Finally, the predictions of molecular shapes by the VSEPR and MO theories are summarized.

Aims

By the end of this chapter you should understand:

- VSEPR theory applied generally to polyatomic molecules
- Valence bond and molecular orbital treatments of molecular shapes
- Why the ammonia molecule is pyramidal
- The bonding in diborane and ethane

6.1 Ammonia, NH_3

6.1.1 VSEPR Theory

The VSEPR theory of the ammonia molecule starts with the nitrogen atom ($2s^2 2p^3$), to which is added the 1s electrons from the three hydrogen atoms, so producing the $2s^2 2p^6$ configuration in the valence shell of the nitrogen atom. The four pairs of σ electrons adopt a tetrahedral distribution to minimize electron pair repulsions. One of the four tetrahedral positions is occupied by a lone pair, the other three being those of the three bonding pairs. The molecular shape should be trigonally pyramidal, with bond angles somewhat smaller than those of a regular tetrahedron, because of the extra repulsion between the lone pair and the bonding pairs (see Section 5.2.1). The molecule is a trigonal pyramid with HNH bond angles of 107°.

6.1.2 Molecular Orbital Theory

The ammonia molecule is a trigonal pyramid, belonging to the C_{3v} point group. The 2s and 2p orbitals of the nitrogen atom and the 1s orbital group combinations of the three hydrogen atoms transform, with respect to the C_{3v} point group, as indicated in Table 6.1.

Table 6.1 Orbitals of the nitrogen and hydrogen atoms of the ammonia molecule

Orbital	C_{3v}
2s(N)	a_1
$2p_x(N) + 2p_y(N)$	e
$2p_z(N)$	a_2
3 × 1s(H)	a_1 + e

Worked Problem

Q Look up the transformation properties of the 2s and 2p orbitals of the nitrogen atom in the character tables of the C_{3v} point group in Appendix 1 to confirm the content of Table 6.1. Carry out the procedure for classifying the 1s orbitals of the three hydrogen atoms as group orbitals in the pyramidal molecule.

A The results of estimating the number of the 1s hydrogen ligand orbitals that are unaffected by the symmetry operations of the C_{3v} point group are:

	E	C_3	σ_v
3 × 1s(H)	3	0	1

This representation of the group orbitals may be converted to its constituent irreducible representations by considering that it is likely that they are identical with the representations to which the 2s and 2p atomic orbitals of the nitrogen atom belong. The subtraction of the characters of the a_1 representation gives the result:

	E	C_3	σ_v
$3 \times 1s(H)$	3	0	1
a_1	1	1	1
$3 \times 1s(H) - a_1$	2	−1	0

The characters in the last row are those of the representation e. The conclusion is that the hydrogen ligand orbitals transform as the sum $a_1 + e$, the same as the 2s and $2p_z$ orbitals of the nitrogen atom, both a_1, and the $2p_x$ and $2p_y$ orbitals, together e.

The 2s and $2p_z$ orbitals of the nitrogen atom have the same symmetry, a_1, and therefore can mix or hybridize. The description of the bonding in ammonia is aided by allowing the two orbitals to mix before constructing the MOs. The process occurs along the z axis, which coincides with the C_3 axis of the molecule. The two linear combinations may be written as in equations 6.1 and 6.2, neither of which contains a normalization factor, nor do they indicate the exact 2s:2p mixtures; the c factors allow for unequal mixing, which is necessary because of the large energy gap between the two atomic orbitals.

$$\phi(2a_1) = c_1\psi(2s) - c_2\psi(2p_z) \tag{6.1}$$

$$\phi(3a_1) = c_2\psi(2s) + c_1\psi(2p_z) \tag{6.2}$$

The two hybrid orbitals are shown pictorially in Figure 6.1. Figure 6.1 also shows the positions of the a_1 group orbital of the three hydrogen atoms. The minus sign in equation 6.1 indicates that the $2p_z$ contribution to $2a_1$ is that which is directed towards the three hydrogen atoms in the ammonia molecule, the $3a_1$ hybrid orbital pointing along the positive direction of the z axis, also shown in Figure 6.1. The $2a_1$ nitrogen orbital may interact with the a_1 group orbital of the three hydrogen atoms to give bonding and anti-bonding combinations, labelled as $2a_1$ and $4a_1$ in the MO diagram shown in Figure 6.2. The $3a_1$ orbital has some 2s content which reduces its energy, but it is essentially non-bonding and the two electrons which it contains are localized along the z axis pointing away from the direction where the hydrogen atoms are located.

There is a doubly degenerate interaction between the e-type orbitals of the nitrogen atom and those of the ligand hydrogen atoms to give the

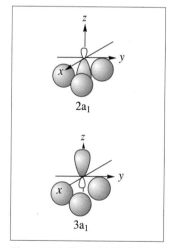

Figure 6.1 The mixing of the 2s and $2p_z$ atomic orbitals of the nitrogen atom in the pyramidal ammonia molecule, and the relationship of the hybrid orbitals with the a_1 group orbital of the three hydrogen atoms

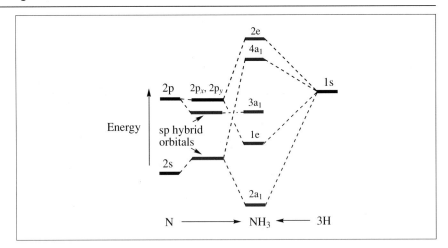

Figure 6.2 The MO diagram of the trigonally pyramidal (C_{3v}) ammonia molecule

molecular orbitals labelled as 1e (bonding) and 2e (anti-bonding). The pyramidal ammonia molecule has the valence electronic configuration $(2a_1)^2(1e)^4(3a_1)^2$, in which the electrons occupying the $2a_1$ and 1e orbitals are bonding, which makes the total bond order 3, shared out between the three supposed bonds between the nitrogen atom and the three ligand hydrogen atoms and gives such bonds single bond status.

Box 6.1 HOMO and LUMO of Ammonia have the Same Symmetry

It is important to notice that the HOMO and LUMO of ammonia have the same symmetry, a_1. This may be contrasted with the hypothetical *planar* form, belonging to the D_{3h} point group, in which the 2s and $2p_z$ orbitals of the nitrogen atom have different symmetries, a_1' and a_2'' respectively (check this is so by using the D_{3h} character table in Appendix 1). This means that they cannot mix. When distortion occurs to give the trigonal pyramid, the HOMO ($1a_2''$) and the LUMO ($3a_1'$) are converted into the $3a_1$ and $4a_1$ orbitals of the pyramidal molecule. This leads to the stabilization expected when this phenomenon occurs. The main stabilization arises when the non-bonding orbital in the planar molecule ($2p_z$) gains some 2s character in becoming the $3a_1$ orbital of the pyramidal molecule. The orbital is still non-bonding, but has a lower energy because it has some 2s character. The stabilization is opposed by the increase in interproton repulsion, so the actual bond angle of 107° is somewhat greater than 90°.

The photoelectron spectrum of the ammonia molecule is consistent with the molecular orbital conclusions: there are three ionization energies at

1050, 1445 and 2605 kJ mol^{-1}, corresponding to the removal of electrons from the 3a$_1$, 1e and 2a$_1$ molecular orbitals, respectively.

6.2 Methane, CH$_4$

6.2.1 VSEPR Theory

The methane molecule is a very important molecule in organic chemistry, the geometry around the tetravalent carbon atom being basic to the understanding of the structure, isomerism and optical activity of a very large number of compounds. It is a tetrahedral molecule belonging to the tetrahedral point group, T_d.

The VSEPR theory assumes that the four electrons from the valence shell of the carbon atom plus the valency electrons from the four hydrogen atoms form four identical electron pairs which, at minimum repulsion, give the observed tetrahedral shape. To rationalize the tetrahedral disposition of four bond-pair orbitals with those of the 2s and three 2p atomic orbitals of the carbon atom, sp^3 hybridization is invoked.

Worked Problem

Q Draw the box diagrams associated with the 'thought experiment' which describes the VSEPR approach to the determination of the methane structure.

A

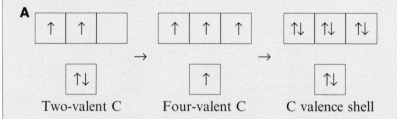

Two-valent C   Four-valent C   C valence shell

The carbon two-valent state is converted to its four-valent state by unpairing the 2s electrons and promoting one of them to the third 2p orbital. The electrons from the four hydrogen atoms (red) make up four pairs of σ bonding electrons, which repel each other to a position of minimum repulsion to give the tetrahedral shape for the methane molecule.

6.2.2 Molecular Orbital Theory

The transformation properties of the orbitals of the carbon atom and the group orbitals of the four hydrogen atoms, with respect to the T_d point group, are given in Table 6.2.

In carrying out VSEPR predictions, bear in mind that the unpairing of electrons and the promotion of electrons to higher energy orbitals, such as p or d, do not necessarily have any connection with the real physical processes that occur when molecules are produced from their elements. Such procedures should be regarded as 'thought experiments' that give the correct shape, but do not necessarily imply the correct bonding.

Table 6.2 Symmetries of the valence orbitals of the carbon atom and the group orbitals of the hydrogen atoms of the methane molecule

Orbital	T_d
2s(C)	a_1
$2p_x(C) + 2p_y(C) + 2p_z(C)$	t_2
$4 \times 1s(H)$	$a_1 + t_2$

The MO diagram for the methane molecule is shown in Figure 6.3. Atomic overlap diagrams for the two sets of bonding MOs are incorporated into Figure 6.3, and emphasize their five-centre nature. In T_d symmetry there is a total match between the two a_1 and two t_2 sets of orbitals (C and H). The electronic configuration of the methane molecule is $(2a_1)^2(1t_2)^6$. All eight electrons occupy bonding orbitals, and give all four bonds single bond status.

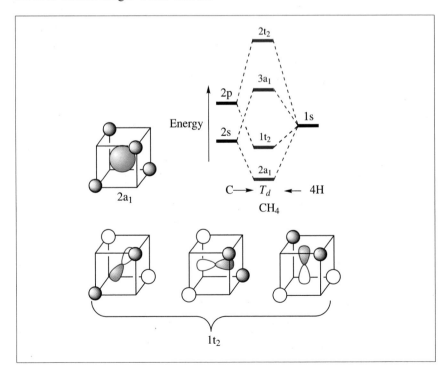

Figure 6.3 The MO diagram of tetrahedral (T_d) methane, and overlap diagrams for the $1a_1$ and $1t_2$ orbitals. The 1s orbital of the carbon atom is not shown, but would be labelled $1a_1$

Worked Problem

Q Use Figure 6.3 to predict the outcome of the addition of an electron to the ammonium ion, NH_4^+.

A The CH_4 and NH_4^+ species are isoelectronic, *i.e.* they possess the same number of electrons. Isoelectronic species usually have the same symmetry and bonding. Figure 6.3 may be used to represent the bonding in the NH_4^+ ion. It would have the electronic configuration $(2a_1)^2(1t_2)^6$. An added electron to produce the neutral molecule NH_4 would occupy the anti-bonding $3a_1$ orbital, which would reduce the bond order of each bond to below one. The unstable molecule would dissociate to give the stable ammonia molecule and a hydrogen atom. The hydrogen atoms so produced would dimerize to give stable dihydrogen molecules:

$$NH_4^+ + e^- \rightarrow NH_4 \rightarrow NH_3 + \tfrac{1}{2}H_2$$

The VSEPR assumption that there are four identical localized electron pair bonds in the four C–H regions, made up from sp^3 hybrid carbon orbitals and the hydrogen 1s orbitals, is not consistent with the experimentally observed photoelectron spectrum. The MO theory is consistent with the two ionizations shown in the photoelectron spectrum of CH_4 and implies that the bonding consists of four electron pairs which occupy the $1a_1$ and $1t_2$ five-centre MOs.

The photoelectron spectrum is shown in Figure 6.4 and shows that

Herzberg (Nobel prize for Chemistry, 1971) commented on the two distinct photoionizations from methane that "this observation illustrates the rather drastic nature of the approximation made in the valence bond treatment of CH_4, in which the 2s and 2p electrons of the carbon atom are considered as degenerate and where this 'degeneracy' is used to form tetrahedral orbitals representing mixtures of 2s and 2p atomic orbitals. The molecular orbital treatment does not have this difficulty".

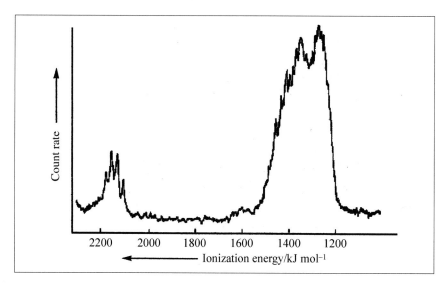

Figure 6.4 A photoelectron spectrum of the methane molecule

there are two ionization energies at 1230 and 2160 kJ mol^{-1}, corresponding to the removal of electrons from the $1t_2$ and $1a_1$ MOs, respectively. The valence electrons in CH_4 are accommodated in two levels of energy rather than the single level implied by the hybridization approach.

6.3   Boron, Nitrogen and Chlorine Trifluorides

Most of the principles by which AB_3 molecules, such as BF_3, NF_3 and ClF_3, are treated have been dealt with in previous sections. For this reason, their treatments are presented in a relatively concise fashion with only additional points being stressed.

6.3.1 Boron Trifluoride

The VSEPR treatment of the shape of the BF_3 molecule depends upon the three σ electron pairs repelling each other to a position of minimum repulsion, giving the molecule its trigonally planar configuration. The molecule belongs to the D_{3h} point group, but experimental observation of the B–F internuclear distance (130 pm) shows it to be smaller than that in the BF_4^- ion (153 pm). Between the same two atoms a shorter distance implies a higher bond order. The VSEPR treatment ignores any possible contribution to the bonding which may be made by the other electrons of the ligand fluorine atoms. It assumes that the fluorine contributions are just the three σ electrons (one electron per fluorine atom).

Although the boron atom only provides three valence electrons, the four valence shell orbitals are occupied additionally by three σ electrons and a pair of π electrons from the fluorine atoms to make up the normal octet of electrons surrounding central atoms of the main group elements. The three canonical forms of the BF_3 molecule are shown in Figure 6.5.

In valence bond terms the bonding consists of three σ bonds and one **dative** or **coordinate** π bond. The latter are electron pair bonds between two atoms in which both electrons originate from one of the participating atoms.

In the MO treatment of the D_{3h} BF_3 molecule the $2p_z$ orbital of the boron atom transforms as an a_2'' representation, as does one of the linear combinations of fluorine $2p_z$ group orbitals (all F–F π bonding). The MO diagram is shown in Figure 6.6.

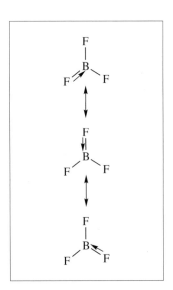

Figure 6.5 The three canonical forms of the BF_3 molecule

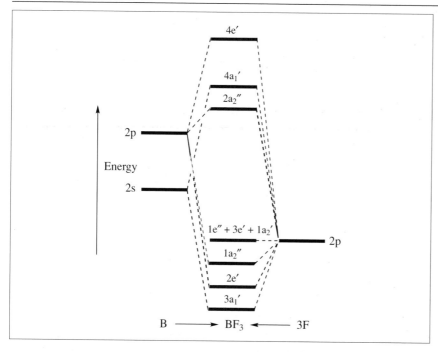

Figure 6.6 A MO diagram for BF_3 with the five non-bonding MOs placed on one level

Worked Problem

Q (i) Take the 2p orbitals of three fluorine atoms pointing towards the centre of an equilateral triangle and classify their combinations in terms of the D_{3h} point group. (ii) Classify the three 2p orbitals which are perpendicular to the trigonal plane.

A (i) The results of estimating the number of the 2p fluorine ligand orbitals in the equilateral plane that are *unaffected* by the symmetry operations of the D_{3h} point group are:

	E	C_3	C_2	σ_h	S_3	σ_v
$3 \times 2p(F)$	3	0	1	3	0	1

This representation of the group orbitals may be converted to its constituent irreducible representations by considering that it is likely that they are identical with the representations to which the 2s and 2p atomic orbitals of the boron atom belong. The subtraction of the characters of the a_1' representation gives the result:

	E	C_3	C_2	σ_h	S_3	σ_v
$3 \times 2p(F)$	3	0	1	3	0	1
a_1'	1	1	1	1	1	1
$3 \times 2p(F) - a_1'$	2	−1	0	2	−1	0

The characters in the last line are those corresponding to the e' representation. That representation is also that of the two 2p orbitals of the boron atom which lie in the xy plane.

(ii) The results of estimating the number of the 2p fluorine ligand orbitals, perpendicular to the trigonal plane, that are unaffected by the symmetry operations of the D_{3h} point group are:

	E	C_3	C_2	σ_h	S_3	σ_v
$3 \times 2p(F)$	3	0	−1	−3	0	1

The minus signs in the C_2 and σ_h columns signify that, although the orbitals have not been moved by the symmetry operations, they have been inverted. This representation of the group orbitals may be converted to its constituent irreducible representations by considering that it is likely that they are identical with the representation to which the $2p_z$ atomic orbital of the boron atom belongs, *i.e.* a_2''. The subtraction of the characters of the a_2'' representation gives the result:

	E	C_3	C_2	σ_h	S_3	σ_v
$3 \times 2p(F)$	3	0	−1	−3	0	1
a_2''	1	1	−1	−1	−1	1
$3 \times 2p(F) - a_2''$	2	−1	0	−2	1	0

The characters in the last line are those corresponding to the e'' representation. The results are incorporated into Figure 6.6.

Worked Problem

Q Why are the B–F bond lengths in BF_3 and BF_4^- different?

A There are three σ bonding orbitals ($3a_1'$ and $2e'$) and a π-type $1a''$ bonding orbital which are all fully occupied and account for the bond order of each B–F linkage being $1\frac{1}{3}$, and is in accordance with the bond length being smaller than in BF_4^- which has a B–F bond order of 1.0. The five non-bonding orbitals are filled.

Box 6.2 Bond Enthalpy Terms

The strengths of the bonds in BF_3, NF_3 and ClF_3 can be estimated from the enthalpies of formation of the compounds, together with the enthalpies of atomization of the constituent elements. The calculations are based on the first law of thermodynamics. The constituent elements are atomized, using the relevant enthalpies of atomization, and the resulting gaseous atoms are considered to form three bonds (element–fluorine), which causes the release of enthalpy to the extent of three times the bond enthalpy term for the particular element–fluorine combination. The overall enthalpy change is then equated with the observed enthalpy of formation of the compound, and the equation solved for the bond enthalpy term. The necessary data are given in Table 6.3.

Table 6.3 Data for bond enthalpy calculations

Element	Standard enthalpy of atomization/ kJ mol^{-1}	Standard enthalpy of formation of EF_3/kJ mol^{-1}
B	590	−1111
N	473	−114
Cl	121	−163
F	79	

The calculation of the bond enthalpy term for boron–fluorine bonds in BF_3 is:

$$E\left(B-F\right) = \frac{1111 + 590 + \left(3 \times 79\right)}{3} = 646 \text{ kJ mol}^{-1}$$

The Mulliken numbering system is used for all MO diagrams; the 1s orbitals of the four atoms in BF_3 and the 2s orbitals of the three fluorine atoms are omitted from the diagram in Figure 6.6.

Worked Problem

Q Carry out the calculations for the N–F and Cl–F bond enthalpy terms.

A The calculation of the bond enthalpy term for nitrogen–fluorine bonds in NF_3 is:

$$E(N-F) = \frac{114 + 473 + (3 \times 79)}{3} = 275 \text{ kJ mol}^{-1}$$

The calculation of the bond enthalpy term for chlorine–fluorine bonds in ClF_3 is:

$$E(Cl-F) = \frac{163 + 121 + (3 \times 79)}{3} = 174 \text{ kJ mol}^{-1}$$

The NF_3 molecule has two extra electrons, compared to BF_3, which would occupy an anti-bonding π-type $2a_2''$ orbital if it had a trigonally planar shape. The ClF_3 molecule with another two electrons would make use of the anti-bonding σ-type $4a_1'$ orbital if the molecule were to be trigonally planar. It is to avoid use of anti-bonding orbitals that they adopt different symmetries.

6.3.2 Nitrogen Trifluoride, NF_3

In VSEPR theory the valence shell configuration of the nitrogen atom in NF_3 is $2s^2 2p^3$, and the addition of the three σ-type electrons from the fluorine atoms gives the octet $2s^2 2p^6$: four pairs of σ electrons which distribute themselves tetrahedrally around the nitrogen atom. The three identical bonding pairs are squeezed together somewhat by the extra repulsion of the lone pair to give a trigonally pyramidal molecule with bond angles slightly less than the regular tetrahedral angle. The observed angle is 104°.

If the NF_3 molecule possessed a trigonally planar structure, its electronic configuration would be that of the BF_3 molecule with the extra pair of electrons occupying the anti-bonding $2a_2''$ (π) orbital. That would cancel out any bonding effect of the two electrons in the bonding $1a_2''$ (π) orbital, and it would not have the extra stability of the π bonding possessed by the BF_3 molecule. For trigonally planar NF_3 the highest energy occupied molecular orbital (HOMO) would be the anti-bonding $2a_2''$ orbital and the LUMO would be the anti-bonding σ-type $4a_1'$ orbital. If a distortion from D_{3h} symmetry to C_{3v} occurs, both orbitals have a_1 symmetry. In that case, they can mix, which leads to the stabilization of the lower energy orbital at the expense of the higher one. In the trigonally pyramidal molecule, the HOMO is non-bonding. The MO diagram shown in Figure 6.7 is constructed by allowing the 2s and $2p_z$ orbitals of the nitrogen atom to hybridize so that one hybrid is directed symmetrically towards the three ligand fluorine atoms and the other hybrid is 180° from that orbital directed away from the bonding region.

This is the same procedure as adopted for the ammonia molecule; see Section 6.1.2.

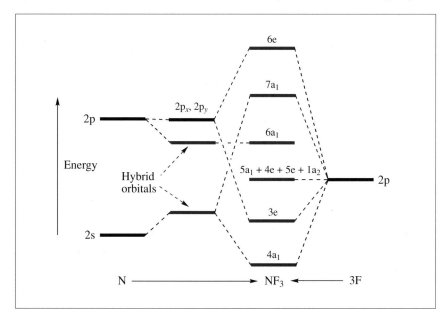

Figure 6.7 A MO diagram for NF_3

The hybrid pointing towards the ligand fluorine atoms and the other two 2p orbitals of the nitrogen atom form the three bonding orbitals ($4a_1$, $3e$) and the corresponding three anti-bonding orbitals ($7a_1$, $6e$). The other six orbitals (derived from the 2p orbitals of the three fluorine atoms), and which are localized on the fluorine atoms, form non-bonding group orbitals with the symmetries and numbering $5a_1$, $4e$, $5e$ and $1a_2$. They are filled in the NF_3 molecule, as is the hybrid orbital directed away from the bonding region, and which has the label $6a_1$, and forms the HOMO. Thus the molecule achieves an N–F bond order of one per bond, and avoids the use of an anti-bonding orbital. The major effect on the energies of the MOs as the symmetry changes from trigonal plane to trigonal pyramid are shown in Figure 6.8. The $2a_2''$ and $4a_1'$ orbitals both adopt a_1 symmetry, and allow the stabilization of the two electrons in the $6a_1$ orbital of the trigonally pyramidal molecule. The ClF_3 molecule with a trigonally pyramidal symmetry would have to use the anti-bonding $5a_1$ orbital and avoids this by adopting a lower symmetry.

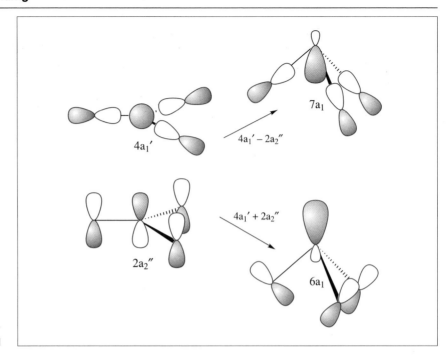

Figure 6.8 A diagram showing the interaction of the 4a$_1$′ and 2a$_2$″ MOs of trigonally planar NF$_3$ permitted by a distortion of the molecule, giving a shape of lower symmetry, *i.e.* trigonally pyramidal

6.3.3 Chlorine Trifluoride, ClF$_3$

The VSEPR treatment of the ClF$_3$ molecule employs a chlorine 3d orbital to contain one of the electron pairs. The chlorine atom is normally $3s^2 3p^5$, and as such is expected to be univalent. To arrange for it to be trivalent one of the paired-up 3p electrons is excited to a 3d level. Then, the three electrons from the ligand fluorine atoms may be added to give the configuration $3s^2 3p^6 3d^2$, making five σ electron pairs, three of which are bonding and two are non-bonding. The basic electron pair distribution is expected to be trigonally bipyramidal, as in dsp^3 hybridization. There are three possible shapes for ClF$_3$ which result from placing the three fluorine atoms in the two different positions in the trigonal bipyramid, either in the trigonal plane or in a position along the C_3 axis of the trigonal bipyramid. The three possible shapes are (1) trigonally planar, with the two lone pairs in apical positions, (2) trigonally pyramidal, with one lone pair in an apical position and the other in the trigonal plane and (3) T-shaped, with both lone pairs in the trigonal plane. There is a general rule which is based upon a detailed consideration of the repulsions between bonding–bonding and bonding–non-bonding and between two non-bonding pairs of electrons. This is that (for the case of five electron pairs) *non-bonding pairs are more stable in the trigonal plane*. Applied to ClF$_3$, this rule implies that the molecule should be T-shaped. The relatively greater repelling effects of the two in-plane non-bonding pairs cause the angle F–Cl–F to be slightly less than 90°. The observed angle is 87°, as is shown in the diagram of Figure 6.9. The VSEPR treatment, incorporating a d orbital, gives the impression that there are three σ bonds of equal strength, but this is not consistent with the different bond lengths of the ClF_3 molecule.

The molecule belongs to the C_{2v} point group. It may be set up with respect to the z axis as shown in Figure 6.10. A MO diagram is shown in Figure 6.11.

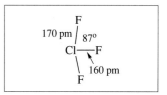

Figure 6.9 The structure of the ClF_3 molecule

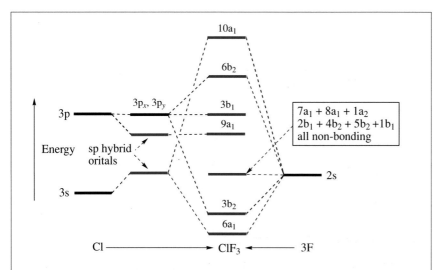

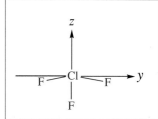

Figure 6.10 The ClF_3 molecule set up in the yz plane

Figure 6.11 A MO diagram for ClF_3

The 3s and $3p_z$ orbitals of the chlorine atom may be hybridized to give one hybrid directed towards the fluorine atom positioned on the z axis, the other pointing in the opposite direction. The hybrid directed towards the fluorine atom can undertake MO formation via the $2p_z$ orbital of the fluorine atom. The only other bonding combination is between the $3p_y$ orbital of the chlorine atom (b_2) and a group orbital of that symmetry constructed from the $2p_y$ orbitals of the fluorine atoms positioned along the y axis. That leaves seven non-bonding orbitals originating on the fluorine atoms and the hybrid orbital pointing away from the bonding region ($6a_1$) along the z axis. That and the $3p_x$ orbital of the chlorine atom ($3b_1$) are also non-bonding. The latter orbital is the HOMO for the molecule. Thus, some bonding is achieved and the use of anti-bonding orbitals avoided, but the overall bonding consists of only two bonding pairs, one localized in the bond between chlorine and the fluorine atom along the z axis and the other forming a 3c2e bond along the y axis.

The symmetries of the d orbitals of the chlorine atom allow some stabilization of the two bonding orbitals, which is consistent with the assumptions made by the VSEPR approach. Both theoretical approaches offer a reason for the deviation from T-shape to an arrow shape for

Box 6.3 The Avoidance of the Use of Anti-bonding Orbitals

The avoidance of the use of anti-bonding orbitals, which would be used in the trigonally planar form of the ClF_3 molecule, is accomplished when the HOMO ($4a_1'$) and LUMO ($4e'$) of the D_{3h} molecule (see Figure 6.6) are transformed into the non-bonding $9a_1$ and anti-bonding $10a_1$ orbitals of the C_{2v} molecule. Additionally, the $2a_2''$ anti-bonding orbital becomes the non-bonding $3b_1$ orbital, and confers more stabilization on the T-shaped molecule.

the ClF_3 molecule. VSEPR theory postulates that there are two lone pairs in the trigonal plane containing the unique Cl–F bond, which repel the two bonding pairs along the y axis. MO theory has the sp hybrid orbital diametrically opposed to the unique Cl–F bond, which can cause the same type of distortion.

Worked Problem

Q Why do the molecules BF_3, NF_3 and ClF_3 have different shapes?

A BF_3 is a trigonal plane in which the inter-ligand repulsions are minimized, and no anti-bonding orbitals are occupied. If NF_3 and ClF_3 were trigonal planar, one and two anti-bonding orbitals, respectively, would have to be used to accommodate electrons. In the trigonally planar NF_3 the HOMO and LUMO are the $2a_2''$ and $4a_1'$ orbitals of Figure 6.6. If a distortion to trigonally planar occurs, these orbitals become the $6a_1$ and $7a_1$ of Figure 6.7. Because they are of the same symmetry they interact so that the occupied orbital is stabilized and this gives overall stability to the pyramidal molecule. In the trigonally planar ClF_3 the HOMO and LUMO are the $4a_1'$ and $4e'$ orbitals of Figure 6.6. If a distortion to the T-shaped C_{2v} symmetry occurs, these orbitals become the $9a_1$ and $10a_1$ of Figure 6.11. Because they are of the same symmetry they interact so that the occupied orbital is stabilized and this gives overall stability to the C_{2v} molecule. The occupied $9a_1$ non-bonding orbital contributes to the molecule having a blunt arrow-head shape rather than a regular T-shape.

6.3.4 Bond Angle Comparisons in Some Group 15 Hydrides and Halides

Table 6.4 contains some bond length and bond angle data for the hydrides, fluorides and chlorides of nitrogen, phosphorus and arsenic. The differences in the electronegativity coefficients (Allred–Rochow) in the compounds are also given.

Table 6.4 Bond length, bond angle and electronegativity coefficient ($\Delta\chi$) data for some EL_3 compounds, where E = N, P or As, and L = H, F or Cl

Molecule	Property	NL_3	PL_3	AsL_3
EH_3	Bond angle	106.6°	93.5°	91.8°
	r(E–H)/pm	101.5	142	152
	r(E–E)/pm	163	207	218
	$\Delta\chi$	0.9	0.1	0
EF_3	Bond angle	102.1°	100°	102°
	r(E–H)/pm	137	154	171
	r(E–E)/pm	213	236	266
	$\Delta\chi$	1.0	2.0	1.9
ECl_3	Bond angle	107.1°	100.1°	98.4°
	(E–H)/pm	176	204	216
	r(E–E)/pm	283	313	327
	$\Delta\chi$	0.3	0.7	0.6

Because the relevant s and p orbitals of phosphorus and arsenic are closer together and at higher general energies than the 2s and 2p orbitals of oxygen, it is not possible to use any of the diagrams of this chapter to discuss their electronic configurations. Some general comments may be made about the trends in bond angles. In the three series of compounds there is very little correlation of bond angle with the size of the ligand atoms. Across the series of EH_3, EF_3 and ECl_3 compounds there is a reduction in the bond angle as the size of the central atom increases. The limit of the size reduction seems to be 90°, corresponding to the full participation of the two p orbitals in the bonding and to the maximization of electronic stabilization. This latter factor overrides the effects of inter-ligand repulsion. There is no noticeable correlation between the physical data and the differences in the electronegativity coefficients of the participating elements in the compounds considered.

From a VSEPR viewpoint it might be expected that the non-bonding pairs would have less effect on reducing the bond angles from the expected regular tetrahedral angle as the size of the central atom increased. It might also be expected that as the ligand atoms become more electronegative they would have a greater share of the bonding electrons and would repel each other more strongly, leading to a larger bond angle. Both predictions from VSEPR theory seem not to hold.

6.4 Ethane, C_2H_6, and Diborane, B_2H_6

The ethane molecule, C_2H_6, can be regarded as the dimer of the free radical CH_3, which has an odd electron. The two odd electrons from two methyl radicals can pair up to make the conventional C–C σ bond to give the covalently saturated ethane molecule. The two methyl groups have rotational freedom with respect to rotation around the C–C axis, although it has been estimated that the difference in energy between the eclipsed and staggered forms of the molecule is 12.6 kJ mol^{-1}, an amount not sufficiently large to offer the possibility of their independent separate existence. The geometry around each carbon atom is expected to be tetrahedral, as indicated from the VSEPR theory applied to four valence shell electron pairs.

Worked Problem

Q Why do compounds of formulae XH_2CCH_2X not have *cis–trans* isomers?

A The C–C bond is a σ bond and is cylindrically symmetric. There is free rotation about such a bond, so there is only one form of the XH_2CCH_2X compounds.

The diborane molecule, B_2H_6, is the simplest of the boron hydrides and can be considered as the dimer of the BH_3 molecule, which has only a transient existence. There are important differences between the diborane molecule and that of ethane. In addition to being a molecule with two 'central' atoms in the VSEPR sense, diborane poses an unusual problem in chemistry. The molecule is described as 'electron deficient' in the sense that there are only 12 valency electrons (three from each B plus one from each H) available for the bonding of the eight atoms. The structure of the molecule is shown in Figure 6.12. The boron atoms are surrounded by four hydrogen atoms, two of which (the bridging atoms) are shared between them. The coordination of the boron atoms is irregularly tetrahedral with HBH angles of 97° for the bridging hydrogen atoms and 121.5° for the terminal angles. The molecule belongs to the D_{2h} point group. The terminal B–H distances are 119 pm and the bridge bonds are all 133 pm long.

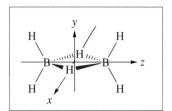

Figure 6.12 The structure of the diborane molecule

The VSEPR theory can be applied to the geometry around both of the boron atoms. The boron atom has the configuration $2s^2 2p^1$ and could be made 'trivalent' by the excitation of one of the 2s electrons to an otherwise vacant 2p orbital: $2s^1 2p^2$. Bearing in mind that the boron is involved in some form of bonding to four hydrogen atoms, the only sensible distribution is that in which the three electrons from the hydrogen atoms [being half the number of 1s(H) electrons available] are placed in the valence shell of the boron atom to give the configuration $2s^2 2p^2 2p^1 2p^1$, with two singly occupied 2p orbitals.

The normal practice of making electron pairs has to be abandoned. The VSEPR argument continues by considering the minimization of the repulsions between the two pairs of electrons and the two unpaired electrons. This gives a distorted tetrahedral distribution with the electron pairs (those involved with the terminal B–H bonding) being further apart than the 'bonds' associated with the single electrons (which participate in the bridging B–H bonding). The two irregular tetrahedra, so formed, join up by sharing the bridging hydrogen atoms to give the observed D_{2h} symmetry of the molecule. Localized bonding could be inferred from this treatment and this leads to the idea of there being four localized one-electron bonds in the bridging between the two boron atoms. The one-electron bond in the H_2^+ molecule-ion is reasonably strong (see Section 3.4). It is a general rule that whenever delocalization can occur, it does. This is because the larger orbitals involved allow for a minimization of interelectronic repulsion. The MO theory as applied so far would classify the orbitals of the boron atoms, and then those of the hydrogen atoms, with respect to their transformation properties in the D_{2h} point group. This may be done but is a very lengthy and complex procedure. With a relatively complex molecule, such as B_2H_6, there is a simpler way of dealing with the bonding which is known as the **molecules within molecule method**.

The 'molecules' within the B_2H_6 molecule, rather than two BH_3 groups, are chosen to be the two BH_2 (terminal) groups and a stretched version of the H_2 molecule. As a separate exercise the bonding in a BH_2 group may be dealt with in exactly the same way as was the H_2O molecule (see Figure 5.12). It belongs to the C_{2v} point group and as such would have the electronic configuration $(1b_2)^2 (2a_1)^2 (3a_1)^2$, with the non-bonding $3a_1$ orbital [it is the $2p_x(B)$ orbital] being singly occupied. There are two important differences from the H_2O case. The 2p–2s energy gap in the boron atom is relatively small ($350\ kJ\ mol^{-1}$) and this allows the 2s(B) and $2p_z(B)$ orbitals (both belonging to a_1) to mix. This has the consequence of making the higher energy $3a_1$ orbital non-bonding. The bonding $2a_1$ is of higher energy than the bonding $1b_2$ orbital because of the very large HBH bond angle which favours the latter orbital.

The two BH_2 groups may be set up so that their z axes are collinear

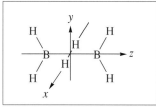

Figure 6.13 Two BH$_2$ groups set up in the yz plane with their C_2 axes collinear with the z axis and with two hydrogen atoms symmetrically placed along the x axis

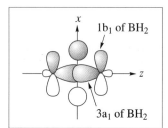

Figure 6.14 The orbitals of two BH$_2$ groups and of two hydrogen atoms used in the formation of the MOs of the bridge bonding in diborane

and coincident with the z axis as in Figure 6.13. Their molecular planes are contained within the yz plane. This ensures that the $1b_1(2p_x(B))$ orbitals of the two BH$_2$ groups are aligned in the x direction and in the xz plane. The two hydrogen atoms responsible for the bridging are situated along the x axis. The two hydrogen group orbitals are used in the construction of those MOs responsible for the cohesion of the three molecules in their formation of diborane. This arrangement of orbitals is shown in Figure 6.14.

In order to construct the MO diagram for the bridging between the two BH$_2$ groups it is essential to re-name the participating orbitals within the framework of the D_{2h} point group to which diborane belongs. This may be done by inspecting the D_{2h} character table and gives the following results:

The orbitals of BH$_2$, $3a_1$ and $1b_1$, form the following linear combinations:

$$\phi(a_g) = \phi(3a_1)_A + \phi(3a_1)_B \tag{6.3}$$

$$\phi(b_{1u}) = \phi(3a_1)_A - \phi(3a_1)_B \tag{6.4}$$

$$\phi(b_{3u}) = \phi(1b_1)_A + \phi(1b_1)_B \tag{6.5}$$

$$\phi(b_{2g}) = \phi(1b_1)_A - \phi(1b_1)_B \tag{6.6}$$

with their new D_{2h} names indicated.

The hydrogen 1s orbitals may be regarded in their usual group forms:

$$\phi(a_g) = \psi(1s)_A + \psi(1s)_B \tag{6.7}$$

$$\phi(b_{3u}) = \psi(1s)_A - \psi(1s)_B \tag{6.8}$$

also with their new D_{2h} names indicated.

The MO diagram for the bridging in diborane may be constructed by the formation of bonding and anti-bonding combinations of the contributing orbitals of the same symmetries. The MO diagram is shown in Figure 6.15. There are two bonding orbitals, of a_g and b_{3u} symmetries, respectively, which contain the four available electrons. The two orbitals are four-centre MOs which allow the minimization of interelectronic repulsion and give the molecule some stabilization compared to the formulation with four localized one-electron bonds.

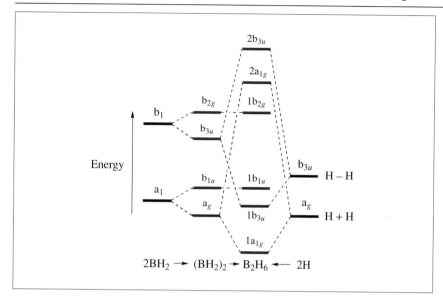

$2BH_2 \rightarrow (BH_2)_2 \rightarrow B_2H_6 \leftarrow 2H$

Figure 6.15 The MO diagram for the bridge bonding in the diborane molecule

6.5   An Overview of Covalency

Two major theories of the covalent bond are described in this book: the main features of valence bond theory are treated in terms of the VSEPR theory of molecular shapes, and MO theory which is based on the symmetry properties of the contributing atomic orbitals. The latter theory is applied qualitatively with MO diagrams being constructed and used to interpret bond orders and bond angles. The problems associated with bond angles are best treated by using the highest symmetry possible for a molecule of a particular stoichiometry.

If the valence electrons can be accommodated in the available bonding and non-bonding orbitals, the high symmetry will be adopted by the molecule. If, however, one or more electrons occupy anti-bonding orbitals in the high-symmetry shape, a distortion occurs to a shape of lower symmetry in which the HOMO (*i.e.* the occupied anti-bonding orbital in high symmetry) is stabilized by interaction with the LUMO so that the electron or electrons in the original HOMO become non-bonding or bonding. The HOMO and LUMO in high symmetry belong to different irreducible representations, but the distortion to lower symmetry allows them to belong to the same irreducible representation (note: this is always the completely symmetric representation of the 'distorted' molecule), so that interaction between them is permitted and leads to the stabilization of the molecule. This mechanism is shown in Figure 6.16.

The predictions of molecular shape by VSEPR theory are summarized in Figure 6.17 for those molecules which have no π bonding. To the left side of Figure 6.17 are the *basic shapes* adopted by molecules which are covalently saturated, *i.e. all* the valence electrons are used in bond for-

The highest symmetry for any molecular shape is the point group with the highest order. The order of a point group is the sum of its component elements. For example, the trigonally planar D_{3h} point group contains the elements E, $2C_3$, $3C_2$, σ_h, $2S_3$ and $3\sigma_v$, making a total of 12. An alternative shape would be the trigonal pyramid belonging to the C_{3v} point group which contains the elements E, $2C_3$ and $3\sigma_v$, a total of 6.

In the examples presented in the text, the distortions from high symmetry which lead to stabilization of the molecular shapes are those in which the HOMO and LUMO of the highly symmetric molecules both become the *completely symmetric representations* of the distorted molecules, *i.e.* a_1. There is a general rule which deals with this phenomenon in a detailed way called the **second-order Jahn–Teller effect**, but that, as they say, is another story.

Figure 6.16 A diagram showing the interaction between the HOMO and LUMO orbitals when a distortion from high symmetry allows them to have the same (a₁) symmetry representation

Energy

LUMO

HOMO

Two molecular orbitals of different symmetries in a molecule of high symmetry

Effect of orbital mixing in a molecule of lower symmetry which allows HOMO and LUMO to belong to the completely symmetric representation (*e.g.* a₁) of the distorted molecule

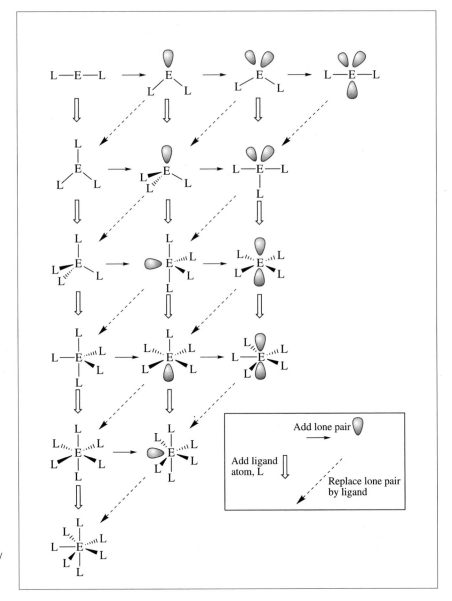

Figure 6.17 A summary of the predictions of molecular shape by the VSEPR theory

mation. Going down the basic shapes, successive E atoms are one group to the right in the Periodic Table. The shapes may be considered as resulting from electron pair repulsion, the basis of VSEPR theory, although the same conclusions arise from considerations of ligand–ligand repulsions. Such ligand–ligand repulsions vary from those between the positive protons in hydride molecules to those between the resultant negative fluorine atoms in fluorides. From left to right across Figure 6.17 are the changes expected when a lone pair of electrons is added to the valence shell, the valency of the atom E remaining constant. Each of the changes represents a shift of two groups to the right in the Periodic Table, *e.g.* Be, C, O and Xe for the top row.

The influence of π bonding on the VSEPR predictions are best summarized by the diagrams of Figure 6.18, which show the changes occurring in a series of intramolecular dehydration reactions. Figures 6.19 and 6.20 show similar dehydration processes occurring when the molecule possesses one and two lone pairs, respectively. Figure 6.21 shows the dehydration reactions of the unusual eight-coordinate $E(OH)_8$ molecule which adopts a square anti-prismatic shape, and eventually reaches the EO_4 oxide. Examples of such molecules are only found in xenon chemistry.

Not all of the hypothetical structures shown in Figures 6.18 and 6.19 correspond to real chemical structures, but most do. A useful exercise would be to consult a book of descriptive inorganic chemistry and identify as many of the structures that represent real molecules.

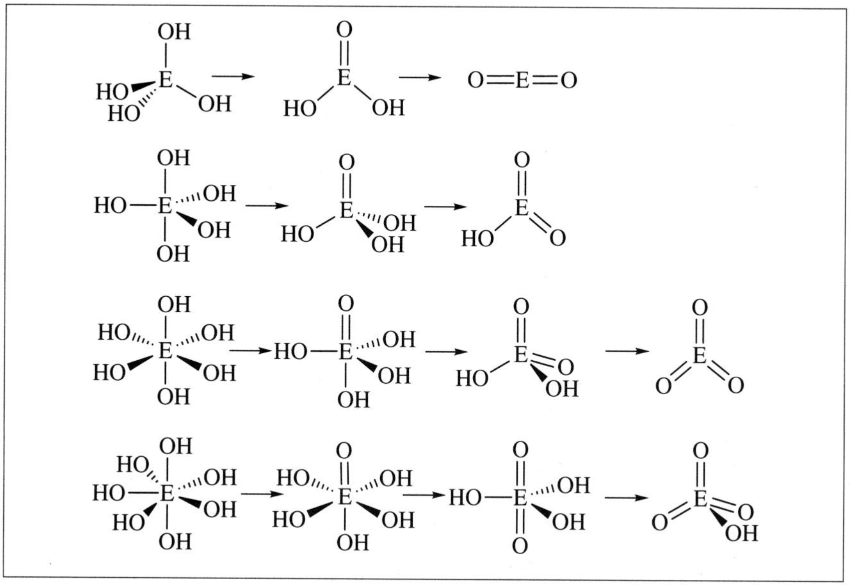

Figure 6.18 Examples of VSEPR theory predictions of molecular shapes in a series of hypothetical dehydration reactions. The element E is covalently saturated

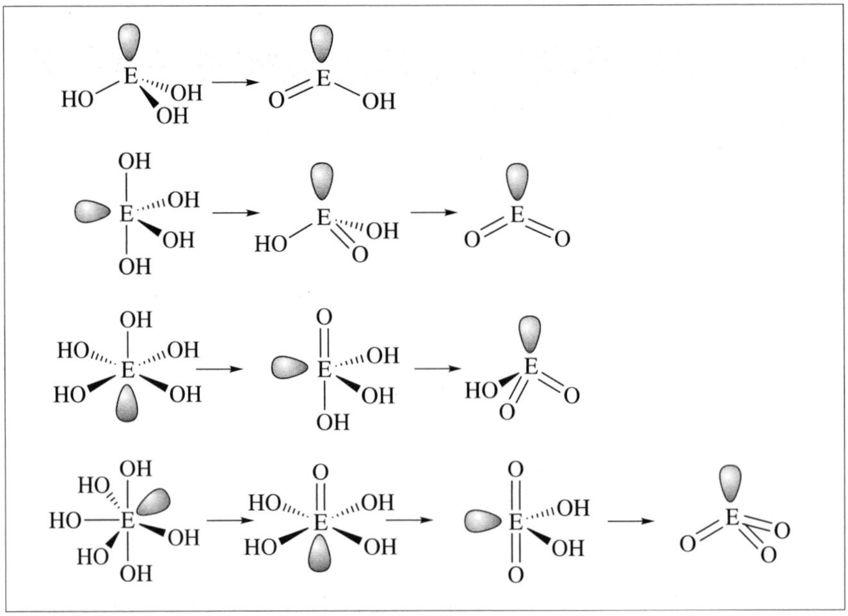

Figure 6.19 Examples of VSEPR theory predictions of molecular shapes in a series of hypothetical dehydration reactions in which element E has one lone pair of electrons

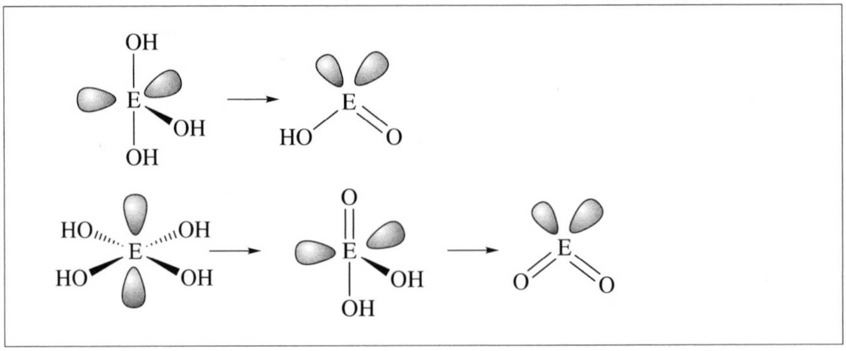

Figure 6.20 Examples of VSEPR theory predictions of molecular shapes in a series of hypothetical dehydration reactions in which element E has two lone pairs of electrons

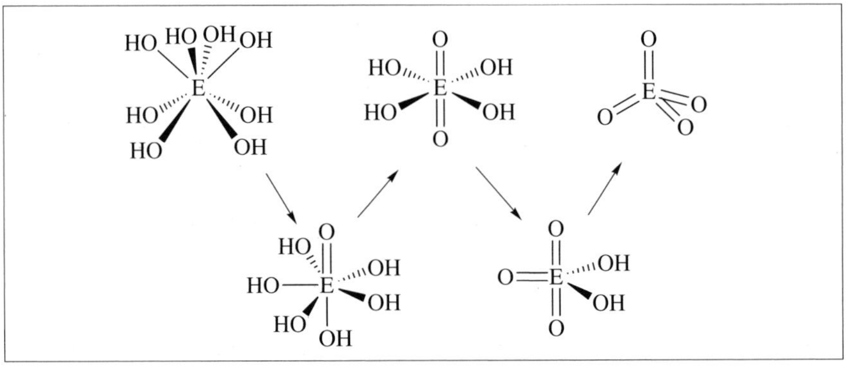

Figure 6.21 The VSEPR theory predictions of molecular shapes in a series of hypothetical dehydration reactions in which element E is eight-coordinate

Summary of Key Points

1. VSEPR/valence bond theory and MO theory were applied to molecules with more than three atoms.

2. The two hydrides NH_3 and CH_4 were discussed extensively, and then a series of molecules with the same stoichiometry but different shapes was dealt with, the trifluorides BF_3, NF_3 and ClF_3.

3. The bonding theories were then applied to two two-centre molecules: B_2H_6 and C_2H_6 were compared.

4. An overview of covalency was presented.

Problems

6.1. The table below contains examples of molecules that possess various total numbers of σ electron pairs, and possess various numbers of bonding electron pairs. For each example, draw out a VSEPR 'thought experiment' diagram and sketch the distribution of the σ electron pairs.

Total number of σ pairs	Number of σ bonding pairs					
	2	3	4	5	6	7
2	CS_2					
3	SO_2	BCl_3				
4	OF_2	NCl_3	$SiCl_4$			
5	ClF_2^-	BrF_3	SCl_4	PF_5		
6			XeF_4	BrF_5	SF_6	
7				XeF_5^-	ClF_6^-	IF_7

Bear in mind that, for the distributions of five and six electron pairs, any lone pairs preferentially occupy the trigonal plane or axial positions, respectively.

6.2. Apply VSEPR and MO theory to the carbonate ion, CO_3^{2-}, and also sketch the canonical forms of the ion which would allow valence bond theory to represent the bonding as accurately as possible. The single bond covalent radii of C and O are 77 and 74 pm, respectively. The C–O bond distance in the carbonate ion is 129 pm.

6.3. Apply the VSEPR and MO theories to SO_3. The S–O distance in the molecule is 141 pm. The single bond covalent radii of S and O are 104 and 74 pm, respectively.

6.4. Hydrogen fluoride reacts with antimony pentasulfide to give the ions H_2F^+ and SbF_6^-. Predict the shapes of the two ions using VSEPR theory and then use the MO approach to describe their bonding.

6.5. Ethyne, ethene and ethane contain 10, 12 and 14 valence electrons, respectively. They are isoelectronic with N_2, O_2 and F_2, respectively. Along both series of molecules the central link becomes progressively weaker and the bond lengths increase. The carbon–carbon bond length increases from 121 pm in ethyne to 133 pm in ethene and to 155 pm in ethane. The corresponding bond enthalpy terms are 837, 612 and 348 kJ mol^{-1}. Data for the diatomic molecules is contained in the text. Discuss these data in terms of the VSEPR and MO treatments of the molecules.

6.6. Silicon tetrafluoride is tetrahedral but SF_4 has a see-saw or butterfly shape. Carry out the VSEPR predictions of these shapes. Consider the MO treatment of SiF_4 as a σ-only case, by using Figure 6.3. Decide if the distortion of a tetrahedral SF_4 to the C_{2v} symmetry is consistent with the HOMO/LUMO stabilization rule, discussed in Chapters 5 and 6.

Further Reading

D. M. P. Mingos, *Essential Trends in Inorganic Chemistry*, Oxford University Press, Oxford, 1998. This book contains discussions of many structural trends.

W. Henderson, *Main Group Chemistry*, Royal Society of Chemistry, Cambridge, 2000. A companion book in the Tutorial Chemistry Texts series.

7

Metallic and Ionic Bonding

About 80% of the elements are solid metals in their standard states at 298 K. Of the metallic elements, only mercury is a liquid at 298 K and at 1 atmosphere pressure (caesium melts at 302 K, gallium at 302.9 K). The non-metallic elements exist as either discrete small molecules, in the solid (S_8), liquid (Br_2) or gaseous (H_2) states, or as extended atomic arrays in the solid state (C as graphite or diamond). The elements of Group 18 are monatomic gases at 298 K.

This chapter consists of two sections, one being a general discussion of the stable forms of the elements, whether they are metals or non-metals, and the reasons for the differences. The theory of the metallic bond is introduced, and related to the electrical conduction properties of the elements. The second section is devoted to a detailed description of the energetics of ionic bond formation. A discussion of the transition from ionic to covalent bonding in solids is also included.

Aims

By the end of this chapter you should understand:

- The structures of the three most common metallic lattices
- The factors which ensure that some elements are metals and others are non-metals
- The essentials of the band theory of the metallic state
- The nature of the ionic bond
- The factors that determine the stoichiometry of ionic compounds
- The factors that determine the transition from ionic to covalent bonding in solids

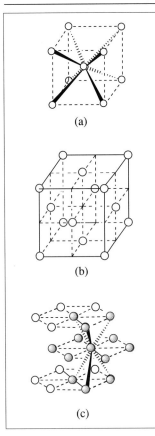

Figure 7.1 Three common crystal lattices adopted by elements: (a) body-centred cubic packing, (b) cubic closest packed (or face-centred cubic) and (c) hexagonal closest packed

7.1 Metal Structures

Metallic character decreases across any period from left to right, and increases down any group, the non-metals being situated towards the top right-hand region of the Periodic Table. The trends in first ionization energies and electronegativity coefficients are opposite to that of metallic character. A large first ionization energy is associated with a large effective nuclear charge, and the atoms in this class are those which can participate in the formation of strong covalent bonds. They have the attracting power to interact with the electrons of another atom in covalent bond formation. Metal atoms, on the contrary, can only form weak covalent bonds with each other, and preferentially exist in the **solid state** with **lattices** in which the **coordination number** (the number of nearest neighbours) of each atom is relatively high (between eight and 14). Collaboration between many atoms is required for metal stability. Diagrams of the three most common metallic lattices are shown in Figure 7.1.

Figure 7.1(a) is the **body-centred cubic lattice** in which the coordination number of each atom is eight. There are six *next nearest neighbours* in the centres of the adjacent cubes only 15% further away, so that the coordination number may be considered to be 14.

Box 7.1 Next Nearest Neighbours

The distance along any side of the body-centred cubic lattice as shown in Figure 7.1(a) is equal to twice the metallic radius of the atom, $2r_M$. The distance between the centre of the atom in the centre of the unit cell and the centre of any atom at the cube corner is $3^{1/3}r_M$. The distance between the centres of two atoms at the centres of adjacent cubes is $2r_M$. This means that any atom in the body-centred cubic arrangement is coordinated by eight atoms at the cube corners with a distance $3^{1/3}r_M$, and six more atoms in the centres of the six adjacent cubes with a distance $2r_M$. The extra six atoms contributing to the coordination number of body-centred atoms are $100 \times (2r_M - 3^{1/3}r_M)/(3^{1/3}r_M) = 15.5\%$ further away from the central atom than the eight nearest neighbours.

The **cubic closest-packed lattice** is shown in Figure 7.1(b), and can also be described as a **face-centred cubic** arrangement in which there is an atom at the centre of each of the six faces of the basic cubic of eight atoms. The coordination number of any particular atom in a cubic closest packed lattice is 12. Considering an atom in one of the face centres of the diagram in Figure 7.1(b), there are four atoms at the corners of the face. Each face of the structure is shared between two such cells, and in the adjacent planes parallel to any shared face there are four atoms. These three sets of four atoms constitute the coordination number of 12.

In the **hexagonally closest packed** arrangement, shown in Figure 7.1(c), the coordination consists of six atoms in the same plane as the atom under consideration plus three atoms from both of the adjacent planes, making a total of 12.

With such high coordination numbers it is quite clear that there can be no possibility of covalency, because there are insufficient numbers of electrons. The difficulty is shown in the case of metallic lithium, with its body-centred cubic structure and coordination number of 14. Each lithium atom has one valency electron and for each atom to participate in 14 covalent bonds is quite impossible.

7.2   Metal or Non-metal?

A discussion of the elementary forms of hydrogen and lithium serves to illustrate the factors which determine their metallic or non-metallic nature. These factors are extendible to elements in general. In a formal sense, the two elements belong to Group 1 of the Periodic Table and are expected to be univalent, as they are. They both have outer electronic configurations which are ns^1, and differ in nuclear charge by two units with lithium having the $1s^2$ 'inner core'. Yet, hydrogen is quite definitely a non-metal and lithium has all the properties of a metal: **metallic lustre** (with lithium this is so for the clean metal surface in the absence of any reactive atmosphere), good **electrical** and **thermal conductivity** in the solid and liquid states, low values of the **photoelectric work function** and relatively small values of the **enthalpy of atomization**. Their electronegativity coefficients, H, 2.2, Li, 0.97, are quite different, the value for lithium placing it beyond doubt in the metallic, electropositive category, and that for hydrogen reflecting the element's three tendencies of either becoming positively or negatively charged (*i.e.* forming H^+ or H^-), or participating in covalent bond formation.

Box 7.2 Work Function

The photoelectric work function is a property of a metallic surface. It is the minimum amount of energy required to cause the removal of an electron from the metallic surface, an experimental procedure usually carried out by irradiating the metal surface (in a vacuum) with photons of sufficient energies to cause the ionization. A plot of the energy of the ejected electrons against the photon energy, extrapolated to zero electron energy, gives the threshold energy of the photon which has just enough energy to cause ionization; this threshold energy is equal to the work function of the particular metal surface.

7.2.1 Elementary Hydrogen

The simplest element exists normally as the dihydrogen molecule, H_2, a strongly covalent diatomic molecule with a bond dissociation energy of 436 kJ mol^{-1}. The high value of the ionization energy of the hydrogen atom is indicative of the electron-attracting power of the bare proton which contributes to the large single bond energy in the dihydrogen molecule.

It is instructive to consider the energetics of formation of larger molecules such as H_4. The hypothetical H_4 molecule could have three possible symmetric structures: tetrahedral, square planar or linear. The results of MO calculations on these molecules are shown in Table 7.1. The nearest H–H distance is assumed to be 100 pm, chosen because it minimized the total energy of the linear form of H_4.

Table 7.1 The calculated energies of some H_4 molecules

Point group	Energy/kJ mol^{-1}		
	Electronic	Nuclear	Total
T_d	−12138	8344	−3794
D_{4h}	−12333	7518	−4815
$D_{\infty h}$	−11524	6017	−5507

The most stable H_4 molecule is the linear $D_{\infty h}$ form, the main factor being the low value of the internuclear repulsion energy. The calculated values for the internuclear repulsion, electronic and total energies for the H_2 molecule (r_{eq} = 74 pm) are 1876, −4807 and −2930 kJ mol^{-1}, respectively. These calculations show that any form of hydrogen which is more than diatomic is less stable than H_2, and that the main reason for this is the nuclear repulsion term. For hydrogen to exist in the metallic form would require tremendous pressure to overcome the internuclear repulsion. It is supposed that such conditions exist in the core of the planet Jupiter, and the presence of metallic hydrogen could explain the magnetic properties of that planet.

7.2.2 Elementary Lithium

The atom of lithium has the outer electronic configuration $2s^1$, and might be expected to form a diatomic molecule, Li_2. The molecule Li_2 does exist in lithium vapour and has a dissociation energy of only 108 kJ mol^{-1}. The enthalpy of atomization of the element is 161 kJ mol^{-1}, which is a measurement of the strength of the bonding in the solid state. To produce two moles of lithium atoms (as would be produced by the dissociation of one mole of dilithium molecules) would require $2 \times 161 = 322$ kJ of energy. From such values it can be seen that solid metallic lithium is more stable than the dilithium molecular form by $(322 - 108)/2 = 107$ kJ mol^{-1}. The energy changes described are shown in the diagram of Figure 7.2, which compares the enthalpies of the metallic element and the diatomic molecule.

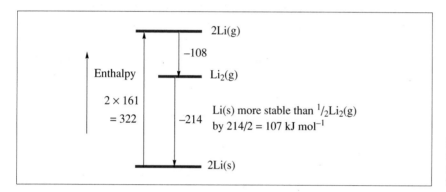

Figure 7.2 A diagram comparing the enthalpies (kJ mol^{-1}) of metallic and gaseous lithium and the Li_2 molecule

The weakness of the covalent bond in dilithium is understandable in terms of the low effective nuclear charge, which allows the 2s orbital to be very diffuse. The addition of an electron to the lithium atom is exothermic only to the extent of 59.8 kJ mol^{-1}, which indicates the weakness of the attraction for the extra electron. By comparison, the exothermicity of electron attachment to the fluorine atom is 333 kJ mol^{-1}. The diffuseness of the 2s orbital of lithium is indicated by the large bond length (267 pm) in the dilithium molecule. The metal exists in the form of a body-centred cubic lattice in which the radius of the lithium atoms is 152 pm: again a very high value, indicative of the low cohesiveness of the metallic structure. The metallic lattice is preferred to the diatomic molecule as the more stable state of lithium.

The high electrical conductivity of lithium (and metals in general) indicates considerable **electron mobility**. This is consistent with the MO treatment of an infinite three-dimensional array of atoms, in which the 2s orbitals are completely **delocalized** over the system with the formation of a **band** of $n/2$ bonding orbitals and $n/2$ anti-bonding orbitals for the n atoms concerned. Figure 7.3 shows a simple representation of the

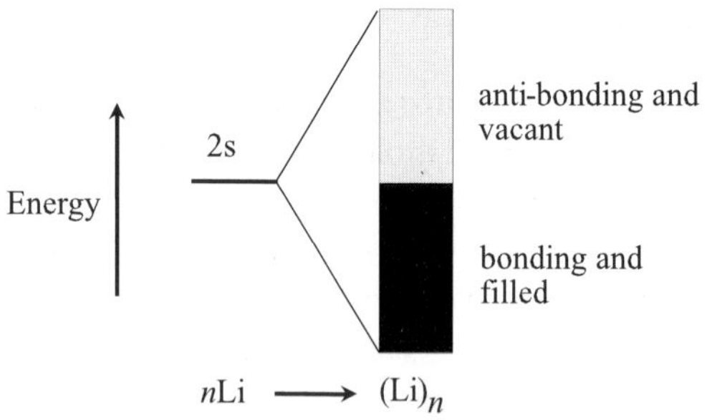

Figure 7.3 The 2s band of lithium metal; the red colouring implies that the lower half is occupied by electrons, the upper shaded half is vacant

2s band of lithium metal. The successive levels (MOs) are so close together that they almost form a continuum of energy. The formation of a band of delocalized MOs allows for the minimization of interelectronic repulsion. The small gaps in energy between adjacent levels in a band of MOs allow a considerable number of the higher ones to be singly occupied. Such single occupation is important in explaining the electrical conduction typical of the metallic state. An electron may join the conduction band at any point of the solid and leave at any other point without having to surmount any significant energy barriers.

The existence of an incompletely filled infinite band of MOs leads to one general definition of the metallic state. The highest filled level in a band is known as the **Fermi level** (equivalent to the HOMO level in small molecules), and when the Fermi level is not at the top of a filled band (as it would be in an insulator or some semiconductors) it leads to the concept of a **Fermi surface** for the infinite array of metal atoms. A metal, then, is that state which has a Fermi surface, one where electron promotion to the nearest higher level is very facile. Apart from electronic transitions within the highest energy band of a metal, there are transitions possible to the next highest otherwise vacant band. In the majority of cases the energy required to cause these inter-band transitions is in the UV region, so that visible light is generally completely reflected from metallic surfaces, giving them their characteristic **'metallic' sheen or lustre**. The apparent colours of some metals, *e.g.* the bluish colour of some steels, are caused by surface layers of oxides which allow interference fringes to be produced. Two metals which have a distinctive 'colour' are copper and gold. In the case of copper the inter-band energy is small enough to allow the absorption of light with frequencies in and around the blue/green part of the visible region to occur, giving the characteristic 'copper' colour. Because of **relativistic effects**, gold has a slightly larger inter-band gap than does copper (although that in silver is larger than both copper and gold), and absorption of the blue/violet part of the visible spectrum gives the metal its golden yellow colour. The colour of gold is discussed in more detail later in this chapter.

7.2.3 Generalizations

For an element to have stable covalent molecules or extended arrays it is essential that it has sufficient valence electrons and that these should experience a high effective nuclear charge, thus making the electron attachment energy sufficiently negative. Elements with electronegativity coefficients greater than about 1.8 fall into this class. Elements with electronegativity coefficients less than 1.8 do not have effective nuclear charges that are sufficiently high to attract electrons from other atoms in order to form covalent bonds, and can achieve thermodynamic stability only by forming delocalized band structures characteristic of metallic bonding. As with any generalization based on an estimate of a property which is difficult to define, such as electronegativity, there are glaring exceptions. For example, silicon ($\chi = 1.7$) and boron ($\chi = 2.0$) are astride the 1.8 suggested borderline between metallic and non-metallic properties, but both are strictly non-metallic from the viewpoint of their not showing metallic conduction of electricity, *i.e.* where the conductivity is high and decreases with increasing temperature. Copper ($\chi = 1.8$) and nickel ($\chi = 1.8$) are on the borderline, but are certainly metals, copper having the next highest electrical conductivity to that of silver, the best conductor. Boron, silicon, germanium, arsenic, antimony, selenium and tellurium are sometimes described as metalloid elements or semi-metals. They show semiconductor properties and have low electrical conductivities which increase as the temperature increases.

The non-metallic elements whose cohesiveness depends upon participation in strong covalent bond formation exist in the following forms:

1. **Diatomic molecules** which may be singly bonded, *e.g.* H_2 and F_2, doubly bonded, *e.g.* O_2, or triply bonded, *e.g.* N_2.
2. **Small polyatomic molecules** involving single covalent bonds between adjacent atoms, *e.g.* P_4 and S_8.
3. **Long chains** which are polymers of small molecules, *e.g.* 'plastic' sulfur composed of polymers of S_8 units with variable chain lengths, and red phosphorus which is a P_4 polymer.
4. **Three-dimensional arrays**, *e.g.* graphite, diamond, boron and silicon. A capacity to be at least three-valent is necessary for a three-dimensional array to be formed.

The individual molecules form crystals in which the cohesive forces are intermolecular, the form of a particular crystal being determined usually by the economy of packing the units together.

Of the elements which participate in **catenation** (chain formation) in their elementary states, only carbon retains the property in its compounds to any great extent. The relatively high strength of the single covalent bond between two carbon atoms gives rise to such a large number, and wide variety, of compounds that it is dealt with in a subsection of the subject called organic chemistry. Boron and silicon chemistry contains some examples of compounds in which catenation is present.

7.3   Metallic Bonding; Band Theory

The band of molecular orbitals formed by the 2s orbitals of the lithium atoms, described above, is half filled by the available electrons. Metallic beryllium, with twice the number of electrons, might be expected to have a full '2s band'. If that were so the material would not exist, since the 'anti-bonding' half of the band would be fully occupied. Metallic beryllium exists because the band of MOs produced from the 2p atomic orbitals overlaps (in terms of energy) the 2s band. This makes possible the **partial filling** of both the 2s and the 2p bands, giving metallic beryllium a greater cohesiveness and a higher electrical conductivity than lithium.

Box 7.3 Cohesiveness of Metal Structures

The cohesiveness of metallic structures (equivalent to their bond strengths) is demonstrated by their enthalpies of atomization, $\Delta_a H^\ominus$. The values for lithium and beryllium are 161 and 321 kJ mol^{-1}, respectively. For comparison, the values for iron and tungsten metals are 418 and 844 kJ mol^{-1}.

The overlapping of the 2s and 2p bands of beryllium, and their partial occupancies, are shown diagrammatically in Figure 7.4. The gap between the 2p and 2s bands of lithium is smaller than that in beryllium, so it is to be expected that some population of the 2p band in lithium occurs and contributes to the high conductivity of the metal. This would also explain why the work function of lithium (280 kJ mol^{-1}) is considerably smaller than the first ionization energy of the isolated atom (513 kJ mol^{-1}). The corresponding data for beryllium are 480 and 900 kJ mol^{-1}, consistent with the partial filling of the 2p band.

The participation of the 2p band in the bonding of metallic beryllium explains the greater cohesiveness (bond strength) of the metal when compared to that of lithium, and also why the enthalpies of atomization and the melting temperatures of the two metals are different, as shown by the data in Table 7.2.

If overlap between bands does not occur, the size of the band gap (between the lowest level of the vacant band and the highest level of the

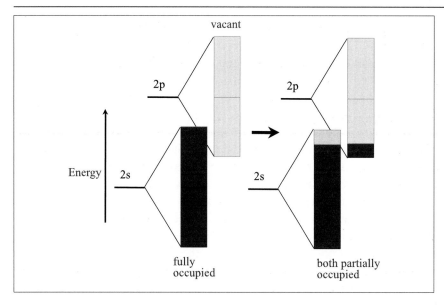

Figure 7.4 The 2s and 2p bands of beryllium metal before (*left-hand side*) and after (*right-hand side*) overlap is allowed to occur; the red colouring indicates filled levels, the shading indicates vacant levels

Table 7.2 Physical properties of lithium and beryllium metals

Property	Lithium	Beryllium
Standard enthapy of atomization/kJ mol^{-1}	161	321
Melting temperature/K	453	1550
Conductivity/S m^{-1}	1.08×10^7	2.5×10^7

filled band) determines whether the element exhibits **insulator properties** or is a **semiconductor**. If the band gap is large the material is an **insulator**. If the band gap is small the material may be a semiconductor. The band gaps in the diamond form of carbon and in silicon (which has the diamond, four-coordinate, structure) are 521 and 106 kJ mol^{-1}, respectively. Diamond is an insulator but silicon is a semiconductor.

7.3.1 Yellow Gold and Liquid Mercury

The colour of gold adds to the attractiveness of the metal, and the liquid state of mercury allows the metal to be used over a wide range of temperatures in thermometers and electrical contact switches. These unusual properties are explicable in terms of relativistic effects. The **relativistic effects** on the 6s orbital are at a maximum in gold and are considerable in mercury.

Relativistic Effects

The general trends across periods and down groups of the Periodic Table are influenced by relativistic effects, which become more serious in the heavier elements ($Z > 55$). This can only be dealt with here in a brief manner.

The theory of relativity indicates that the mass of a particle is dependent upon its velocity. The effect is only noticeable when the particle velocity approaches that of light itself. The average velocity of a 1s electron is proportional to the atomic number (Z) of the element. The relativistic effect exhibits a maximum in gold and mercury, in which atoms the 6s orbital is being filled. The mass of a 1s electron in a gold atom is some 23% greater than the rest mass. Since the radius of the 1s orbital is inversely proportional to the mass of the electron, the radius of the orbital is reduced by 23% compared to that of the non-relativistic radius. This **s-orbital contraction** affects the radii of all the other orbitals in the atom up to, and including, the outermost orbitals. The s orbitals contract, the p orbitals also contract, but the more diffuse d and f orbitals can become more diffuse as electrons in the contracted s and p orbitals offer a greater degree of shielding to any electrons in the d and f orbitals.

The effects are observable by a comparison of the metallic radii, the first three ionization energies and the first electron attachment energy of the Group 11 elements as shown in Table 7.3, and the metallic radii and the first two ionization energies of the elements of Group 12 (Zn, Cd and Hg) as given in Table 7.4.

Table 7.3 The metallic radii, first three ionization energies and first electron attachment energies of the Group 11 elements (energies in kJ mol^{-1})

Atom	r/pm	I_1	I_2	I_3	E
Cu	128	745	1960	3550	−119
Ag	144	732	2070	3360	−126
Au	144	891	1980	2940	−223

Table 7.4 The metallic radii and first two ionization energies of the Group 12 elements

Atom	r/pm	I_1/kJ mol^{-1}	I_2/kJ mol^{-1}
Zn	133	908	1730
Cd	149	866	630
Hg	152	1010	1810

In gold the 6s band is stabilized, making the Fermi level lower than expected, and the 5d level is destabilized. This makes electronic transitions to the d band from the 'Fermi band' have energies that correspond to the blue/violet region of the visible spectrum, so that the metal reflects the red/yellow frequencies (their photon energies are not sufficiently high to cause transitions to the 5d band). The majority of metals have transitions into their 'Fermi bands' which lie in the ultraviolet region of the spectrum, and consequently they reflect white light with very little selective absorption so that sensations of colour are not produced. The stabilization of the 6s orbital in gold is responsible for the very large negative value of the electron attachment energy; only the halogens and platinum have even more negative values.

The mercury atom is smaller than expected from the zinc–cadmium trend and is more difficult to ionize than the lighter atoms. In consequence the metal–metal bonding in mercury is relatively poor, resulting in the element being a liquid in its standard state. This almost Group 18 behaviour of mercury may be compared to that of a real Group 18 element, xenon, which has first and second ionization energies of 1170 and 2050 kJ mol^{-1}.

7.4   Ionic Bonding

The interaction between any two atomic orbitals is at a maximum if the energy difference between them is zero and they have mutually compatible symmetry properties. A difference in energy between the two participating orbitals causes the bonding orbital to have a majority contribution from the lower of the two atomic orbitals. Likewise, the anti-bonding orbital contains a majority contribution from the higher of the two atomic orbitals. In a molecule such as HF the bonding orbital has a major contribution from the fluorine 2p orbital and in consequence the two bonding electrons are unequally shared between the two nuclei, which results in the molecule being **dipolar**.

See the treatment of HF in Section 4.3.4.

The difference in energies between the 3s valence electron of the sodium atom and that of the 2p level in the fluorine atom is so great that the bonding orbital is virtually identical to the 2p(F) orbital, and the anti-bonding orbital is identical to the 3s orbital of the sodium atom. If a bond were formed it would be equivalent to the transfer of an electron from the sodium atom to the fluorine atom, causing the production of the two ions, Na^+ and F^-. A detailed consideration of this specific case is used to explain the nature of **ionic bonding**.

The **standard enthalpy change** for the reaction:

$$Na(s) + \tfrac{1}{2}F_2(g) \rightarrow Na^+(g) + F^-(g) \qquad (7.1)$$

may be estimated as the sum of the standard enthalpy changes of the following four component reactions:

1. $Na(s) \rightarrow Na^+(g)$, the atomization of sodium, $\Delta_{atom}H^{\ominus}(Na) = 109$ kJ mol^{-1}.
2. $\frac{1}{2}F_2(g) \rightarrow F(g)$, the atomization of fluorine, $\frac{1}{2}D(F_2,g) = 79$ kJ mol^{-1}, where $D(F_2,g)$ is the **bond dissociation enthalpy** of the fluorine molecule.
3. $Na(g) \rightarrow Na^+(g)$, the ionization of sodium, $I_1(Na) = 500$ kJ mol^{-1}, where I_1 is the first ionization enthalpy of the sodium atom (*i.e.* the first ionization energy plus $\frac{5}{2}RT$, $494 + 6 = 500$ kJ mol^{-1}).
4. $F(g) \rightarrow F^-(g)$, the attachment of an electron to the fluorine atom, $E_{ea}(F) = -339$ kJ mol^{-1}, where $E_{ea}(F)$ is the electron attachment enthalpy of the fluorine atom (*i.e.* the first electron attachment energy minus $\frac{5}{2}RT$, $-333 - 6 = -339$ kJ mol^{-1}).

The thermodynamic exactitude of allowing for the differences between internal energy changes and changes of enthalpy is not very important as the two 'corrections' cancel out.

The standard enthalpy change for equation (7.1) is given by:

$$\Delta H^{\ominus} = \Delta_{atom}H^{\ominus}(Na) + \tfrac{1}{2}D(F_2,g) + I_1(Na) + E_{ea}(F)$$
$$= 109 + 79 + 500 - 339 = 349 \text{ kJ mol}^{-1}$$

The reaction is considerably endothermic and would not be feasible (changes of **entropy** are excluded from the calculation, but would not alter the conclusion).

The above calculation applies to independent sodium and fluoride ions, and does not take into account the electrostatic attraction between the oppositely charged ions, nor the repulsive force which operates at small interionic distances. In the crystal of NaF the distance of nearest approach of the sodium and fluoride ions is 231 pm, and Coulomb's law may be used to calculate the energy of stabilization due to electrostatic attraction between individual ion pairs:

$$E(Na^+F^-) = -\frac{N_A e^2}{4\pi\varepsilon_0 r} \tag{7.2}$$

where N_A is the Avogadro constant, e the electronic charge, ε_0 the vacuum permittivity and r the interionic distance. Equation 7.2 gives a value of $E(Na^+F^-)$ of -601 kJ mol^{-1}. As a **change of enthalpy** this result has to have 6 kJ mol^{-1} subtracted from it to give -607 kJ mol^{-1}.

The standard enthalpy of formation of the substance Na^+F^- with **discrete ion pairs**, but with no interaction between pairs, is then calculated to be:

$$\Delta_f H^{\ominus}(Na^+F^-,g) = 349 - 607 = -258 \text{ kJ mol}^{-1}$$

This calculation is still hypothetical, in that the actual substance formed when sodium metal reacts with difluorine is solid sodium fluoride, and the standard enthalpy of its formation is -569 kJ mol^{-1}. The actual substance is 311 kJ mol^{-1} more stable than the hypothetical substance consisting of ion pairs, Na$^+$F$^-$(g), described above. The added stability of the observed solid compound arises from the long-range interactions of all the positive Na$^+$ ions and negative F$^-$ ions in the solid **lattice** which forms the structure of **crystalline** sodium fluoride. The ionic arrangement is shown in Figure 7.5. Each Na$^+$ ion is octahedrally surrounded (*i.e.* coordinated) by six fluoride ions, and the fluoride ions are similarly coordinated by six sodium ions. The **coordination numbers** of both kinds of ion are six.

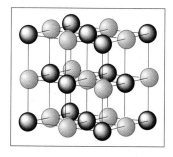

Figure 7.5 The structure of the sodium fluoride crystal

The overall stability of the NaF lattice is represented by the resultant of the many stabilizing attractions (Na$^+$–F$^-$) and destabilizing repulsions (Na$^+$–Na$^+$ and F$^-$–F$^-$), which amount to a stabilization which is 1.74756 times that of the interaction between the individual Na$^+$–F$^-$ ion pairs. The factor, 1.74756, is the **Madelung constant**, M, for the particular lattice arrangement and arises from the forces experienced by each ion. These are composed of six attractions at a distance r, 12 repulsions at a distance $2^{1/2}r$, eight attractions at a distance $3^{1/2}r$, six repulsions at a distance of $4^{1/2}r$, 24 attractions at a distance $5^{1/2}r$, and so on.

The series $6 - 12/2^{1/2} + 8/3^{1/2} - 6/4^{1/2} + 24/5^{1/2} - \ldots$ eventually becomes convergent and gives the value for the Madelung constant for the *sodium chloride lattice* (the standard description of lattices which have the same form as that adopted by sodium fluoride). The values of Madelung constants for some common crystal lattices are given in Table 7.5.

Table 7.5 Values of the Madelung constants for some common lattices

Lattice type	Madelung constant
NaCl	1.74756
CsCl	1.76267
CaF$_2$ (fluorite)	2.51939
TiO$_2$ (rutile)	2.408
ZnS.(blende, sphalerite)	1.63806
ZnS (wurtzite)	1.64132
Al$_2$O$_3$ (corundum)	4.17186

The **electrostatic contribution** to the **lattice energy**, L, for the sodium fluoride arrangement (the energy required to form gas phase ions from the solid crystalline lattice) is the value of the change in internal energy (*i.e.* ΔU) for the reaction:

$$Na^+F^-(s) \rightarrow Na^+(g) + F^-(g) \qquad (7.3)$$

Lattice energies (sometimes defined as the energy released when the lattice is formed from its constituent gaseous ions) are usually tabulated in data books as positive quantities and are internal energy changes at 0 K. The internal energy change corresponding to the formation of a lattice when the gaseous ions condense to produce the crystalline lattice (the reverse of equation 7.3) is given by the equation:

$$\Delta U = -\frac{MN_A e^2}{4\pi\varepsilon_0 r} \qquad (7.4)$$

where M is the Madelung constant and the other quantities are the same as given in equation 7.2.

In addition to the Coulombic forces, there is a repulsive force which operates at short distances between ions as a result of the overlapping of filled orbitals, potentially a violation of the Pauli exclusion principle. This repulsive force may be represented by the equation:

$$E_{\text{repulsion}} = \frac{B}{r^n} \qquad (7.5)$$

where B is a proportionality factor. The full expression for the change in internal energy becomes:

$$\Delta U = -\frac{MN_A e^2}{4\pi\varepsilon_0 r} + \frac{B}{r^n} \qquad (7.6)$$

When r is the equilibrium interionic distance r_{eq}, the differential of equation 7.6 is equal to zero, corresponding to an energy minimum:

$$\frac{\mathrm{d}\Delta U}{\mathrm{d}r} = 0 = \frac{MN_A e^2}{4\pi\varepsilon_0 r_{eq}^2} - \frac{nB}{r_{eq}^{n+1}} \qquad (7.7)$$

which gives an expression for B:

$$B = \frac{MN_A e^2 r_{eq}^{n-1}}{4\pi\varepsilon_0 n} \qquad (7.8)$$

Substitution of this value into equation 7.6 with $r = r_{eq}$ gives:

$$\Delta U = -\frac{MN_A |z^+ z^-| e^2}{4\pi\varepsilon_0 r_{eq}}\left[1 - \frac{1}{n}\right] \qquad (7.9)$$

which is known as the **Born–Landé equation**. The *positive value* of the product of the ionic charges, $|z^+z^-|$, is included in the equation to take into account any charges which differ from ± 1, so that the equation may be applied to any lattice provided that the appropriate value of the Madelung constant is used. The lattice energy, as defined above, has the value $-\Delta U$. The **enthalpy change of lattice formation** (for two gaseous ions forming a lattice) is given by:

$$\Delta_{\text{latt}} H^{\ominus} = \Delta U - 6 \text{ kJ mol}^{-1} \tag{7.10}$$

The values of the **Born exponent**, n, for various crystal structures are estimated from compressibility data. The values recommended for use with various ion configurations are shown in Table 7.6.

Table 7.6 Some values of the Born exponent, n

Ion configuration	Value of n
He	5
Ne	7
Ar, Cu^+	9
Kr, Ag^+	10
Xe, Au^+	12

For a crystal with mixed ion types the average value of n should be used in the calculation of lattice energy. Using the value $n = 7$ for sodium fluoride with r_{eq} equal to 231 pm gives a value for L of 901 kJ mol^{-1}. The condensation of gaseous sodium and fluoride ions to give a mole of sodium fluoride would have an enthalpy change of $-901 - 6 = -907$ kJ mol^{-1}.

It is now possible to produce a theoretical value for the standard enthalpy of formation of sodium fluoride based upon the ionic model described above. The necessary equation is produced by adding together equations 7.1 and 7.3 to give:

$$\text{Na(s)} + \tfrac{1}{2}\text{F}_2 \rightarrow \text{Na}^+\text{F}^-\text{(s)} \tag{7.11}$$

and the calculated value for $\Delta_f H^{\ominus}(\text{Na}^+\text{F}^-)$ is $349 - 907 = -558$ kJ mol^{-1}, which is fairly close ($\sim 98\%$) to the observed value of -569 kJ mol^{-1} and gives considerable respectability to the theory. Figure 7.6 contains all the energy factors that contribute to the overall production of the exothermicity and consequent stability of the sodium fluoride lattice.

In general terms, the feasibility of ionic bond production depends upon the $\Delta_f H^{\ominus}$ value for the compound MX_n, the entropy of formation

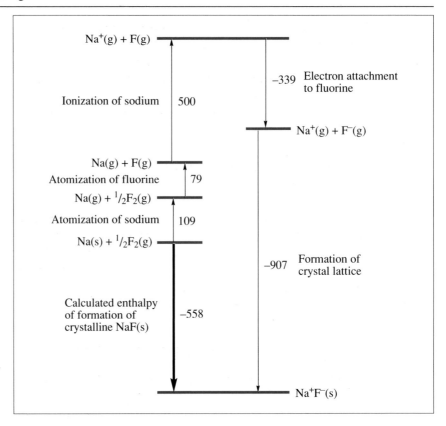

Figure 7.6 An enthalpy (kJ mol⁻¹) diagram for the formation of the sodium fluoride lattice

being of minor importance. The value of $\Delta_f H^{\ominus}$ may be estimated from known thermodynamic quantities and the calculated value for the lattice energy:

$$\Delta_f H^{\ominus} = \Delta_{atom} H^{\ominus}(M) + n\Delta_{atom} H^{\ominus}(X) + \Sigma I(M) + nE_{ea}(X) + \Delta_{latt} H^{\ominus}(MX_n)$$
(7.12)

the enthalpy of atomization of the metal, $\Delta_{atom} H^{\ominus}(M)$, sometimes known as the **sublimation enthalpy**, and that of the non-metal, $\Delta_{atom} H^{\ominus}(X)$, the enthalpy change required to produce the gas phase atom from the element in its standard state, incorporating the change of enthalpy required to produce the element in the gas phase and the enthalpy of dissociation of the molecule. The ionization energy term is the appropriate sum of the first n ionization energies to produce the required n-positive ion. The n electrons are used to produce either n singly negative ions or $n/2$ doubly negative ions and the $E_{ea}(X)$ term has to be modified accordingly. The $\Delta_{latt} H^{\ominus}(MX_n)$ value is calculated, using the Born–Landé equation (7.9) for the appropriate ionic arrangement (if known, otherwise appropriate guesses have to be made) and corrected for the transformation of an internal energy change to a change of enthalpy.

The first three terms of equation 7.12 are positive and can only contribute to the feasibility of ionic bond production by being *minimized*. The last two terms are negative for uni-negative ions (E_{ea} represents energy released), and for ionic bond feasibility should be maximized. From such considerations it is clear that ionic bond formation will be satisfactory for very electropositive and easily atomized metals with very electronegative and easily atomized non-metals.

The lattice energy term may be increased, for ions with charges, z^+ and z^-, by the positive value of the product of the charges, $|z^+z^-|$. To increase the cation charge involves the expense of an increase in $\Sigma I(M)$ and, although there is some pay-back in the production of more anions (a larger E_{ea}), there are severe limits to such activity. This is because the successive ionization energies of an atom increase more rapidly than does the lattice energy with increasing cation charge. The stoichiometry of the ionic compound produced from two elements is determined by the values of the five terms of equation 7.12, which minimize the value of the enthalpy of formation of the compound.

Worked Problem

Q Why do ionic compounds dissociate into atoms in the gas phase?

A Because the gaseous atomic states are more stable than the gaseous ions. The conversion described by the equation:

$$M^{n+}(g) + nX^-(g) \rightarrow M(g) + nX(g) \qquad (7.13)$$

has an enthalpy change given by:

$$\Delta_r H^{\ominus} = -\Sigma I(M) - nE_{ea}(X) \qquad (7.14)$$

the difference between the sum of the n ionization energies of M (released) and the n values of the electron attachment energy of X (absorbed). The values of ionization energies and electron attachment energies are such that the enthalpy change for the conversion of gaseous ions into gaseous atoms is always exothermic. An example of this occurs when salted water boils over on a gas flame. The characteristic yellow sodium emission is observed and that depends upon the 3p $\rightarrow$ 3s transition in the sodium *atom*.

Calculations may be made for hypothetical compounds such as NaF_2. Assuming that the compound contains Na^{2+} ions, the standard enthalpy of formation is given by the equation:

In the calculations of enthalpies of formation of ionic compounds, the differences from the accepted experimental values (which are very accurate) are probably due to the essential simplicity of the model, rather than having any other significance. If large discrepancies are found between experimental and calculated quantities, this probably means that the model is in error and that the compounds have a considerable covalent character. In the Born–Landé equation (and other similar equations) the main uncertainty lies with the n value, but that term only accounts for roughly 10% of the value of the lattice energy. Deviations from the experimental values for calculated enthalpies of formation greater than 10% usually mean that the compound is considerably covalent.

$$\Delta_f H^{\ominus}(NaF_2, ionic) = \Delta_{at} H^{\ominus}(Na) + \Delta_{at} H^{\ominus}(F_2) + I_1(Na) +$$
$$I_2(Na) + 2E_{ea}(X) + \Delta_{latt} H^{\ominus}(NaF_2) \tag{7.15}$$

The numerical values of the terms in equation 7.15 are, respectively, 108, 158, 496, 4562, 2×-339 and (assuming the hypothetical Na^{2+} ion to have a radius of 65 pm, as does the real Mg^{2+} ion, and using a Madelung constant of 2.381, appropriate for the MgF_2 lattice) 2821 kJ mol^{-1}, which gives a value of $+1825$ kJ mol^{-1} for the standard enthalpy of formation of ionic NaF_2, a compound which would, therefore, not be expected to exist. The reason for that is the very large value of the second ionization energy of the sodium atom, the electron being removed from one of the very stable 2p orbitals.

Worked Problem

Q Calculate the standard enthalpies of formation for the compounds (i) sodium chloride and (ii) potassium iodide. The interionic distances in the compounds are 282 and 353 pm, respectively. Compare your answers with the accepted experimental values for these quantities, which are -411 and -327.6 kJ mol^{-1}, respectively.

A The lattice enthalpy of sodium chloride is given by equation 7.9 by using the value of the Madelung constant from Table 7.5 and the value of the Born exponent from Table 7.6 of 8 (the mean value of those given for the two ions). An amount of 6 kJ mol^{-1} has to be subtracted to allow for the conversion of the internal energy change to an enthalpy change. For sodium chloride the enthalpy change corresponding to the formation of the crystalline solid is $-753 - 6 = -759$ kJ mol^{-1}.

Using equation 7.12 with the appropriate values for sodium chloride gives:

$$\Delta_f H^{\ominus}(NaCl, s) = \Delta_{atom} H^{\ominus}(Na) + \Delta_{atom} H^{\ominus}(Cl) + I_1(Na) + E_{ea}(X) +$$
$$\Delta_{latt} H^{\ominus}(NaCl)$$
$$= 109 + 121 + 500 - 355 - 759 = -384 \text{ kJ mol}^{-1}$$

The calculated value is 93% of the accepted value.

(ii) The calculation of the standard enthalpy of formation of KI(s) is similar to that of NaCl(s) except that the Born exponent used is the mean of the values 9 for K^+ and 12 for I^-, *i.e.* 10.5. This gives the enthalpy of formation of the lattice as $-622 - 6 = -628$ kJ mol^{-1}. Using equation 7.12 with the appropriate values for potassium iodide gives:

$$\Delta_f H^{\ominus}(\text{KI,s}) = 90 + 107 + 424 - 301 - 628 = -308 \text{ kJ mol}^{-1}$$

The calculated value is 94% of the accepted value.
Both answers are consistent with the ionic model for these compounds.

7.5 Ionic–Covalent Transition

Both ionic bonding and covalent bonding are extreme descriptions of the real state of affairs in elements and their compounds, and although there are many examples of almost pure, *i.e.* 99%, covalent bonding, there are very few so-called ionic compounds in which there is complete transfer of one or more electrons from the cationic element to the anionic one. This ionic–covalent transition in which an initially purely ionic compound acquires some covalent character is reasonably easy to deal with qualitatively, but is very difficult to quantify. Fajans stated four rules which are helpful in deciding which of two compounds has the most covalency. In this respect the starting point is complete electron transfer in the compounds considered. The positively charged cations are then considered to exert a polarizing effect on the negatively charged anions such that any polarization represents a clawing-back of the transferred electron(s), so that there is some electron sharing between the atoms concerned which may be interpreted as partial covalency. Figure 7.7 shows diagrammatically what is meant by polarization of an anion; the distortion from the spherical distribution of the anionic electrons is a representation of the covalency.

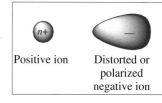

Positive ion   Distorted or polarized negative ion

Figure 7.7 A diagram illustrating the polarization of an ion

Fajans' rules state that the extent of polarization of an anion or the partial covalent character of an otherwise 100% ionic compound is increased by:

1. A high charge of the cation and/or anion
2. A small cation
3. A large anion
4. A cation with a Group 18 gas electronic configuration

The reasoning underlying these rules is dependent upon the **charge density** of the cation, the charge density being the ratio of charge:surface area. The first three rules are straightforward electrostatics. Comparing the anhydrous chlorides of sodium, magnesium and aluminium it is expected that the last compound would have the greatest partial covalency, be the least ionic, and that the sodium compound would be the most ionic.

The fourth rule is an indication that a sub-shell of d electrons is a poorer shield of the nuclear charge than an octet of $s^2 p^6$ electrons, and that a cation with a d shield will exert a greater polarizing effect than one with an s–p shield. Comparing copper(I) chloride with sodium chloride is the classical example of this rule. Cu^I has the electronic configuration $3d^{10}$, whereas Na^+ has the [Ne] core configuration. The ions are similar in size, their ionic radii being 96 pm for Cu^+ and 102 pm for Na^+, but the melting temperatures of their chlorides are 430 °C for CuCl and 801 °C for NaCl, giving some indication that the copper compound is more covalent.

Worked Problem

Q In the supposedly ionic compounds MX and NX_2, where M and N are metallic elements and X is a univalent non-metal, which compound would you expect to have a greater degree of covalency?

A The element N would have a double charge in the compound NX_2, which would give it a higher polarizing power than the singly charges element M in the compound MX. The compound NX_2 would have a greater degree of covalency.

Summary of Key Points

1. The structures of metals were described in terms of the three most common metallic lattices.

2. Metal bonding theory was introduced.

3. The factors which determine whether elements are metals or non-metals were discussed.

4. The nature of ionic bonding was dealt with in detail.

5. The Born–Landé equation for the lattice energy of an ionic crystal was derived.

6. The factors that determine the stoichiometry of ionic compounds were discussed.

7. The transition from ionic character to covalent character was discussed.

Problems

1. The crystal of CaO assumes the NaCl structure, the closest approach between positive and negative ions being 239 pm. The enthalpy of atomization of calcium metal is 193 kJ mol^{-1}; that of dioxygen is 248 kJ mol^{-1} (of oxygen atoms). The two electron attachment enthalpies of oxygen are -148 and $+838$ kJ mol^{-1}, respectively. The first and second successive ionization enthalpies of calcium are 596 and 1156 kJ mol^{-1}, respectively. Use the data to calculate a value for the standard enthalpy of formation of calcium oxide and compare the result with the experimental value of -635 kJ mol^{-1}.

2. In the crystal of ZnS (sphalerite structure, not dealt with in the text) the closest approach between positive and negative ions is 258 pm. The enthalpy of atomization of zinc metal is 130 kJ mol^{-1} and that of elementary sulfur is 223 kJ mol^{-1}. The two electron attachment enthalpies of sulfur are -206 and $+526$ kJ mol^{-1}, respectively. The first and second successive ionization enthalpies of zinc are 914 and 1736 kJ mol^{-1}, respectively. Use the data to calculate a value for the standard enthalpy of formation of zinc sulfide and compare the result with the experimental value of -203 kJ mol^{-1}. How would you interpret the discrepancy?

3. The elements nitrogen and phosphorus of Group 15 have standard states composed of the discrete molecular species N_2 and P_4. The single–single and triple–triple bond energy terms for nitrogen and phosphorus are 163, 944, 172 and 481 kJ mol^{-1}, respectively. Calculate the enthalpy changes for the reactions $2N_2 \rightarrow N_4$ and $2P_2 \rightarrow P_4$ to show the relative stabilities of the two atomicities for the two elements. Assume that N_4 is tetrahedral, as is P_4.

References

P. Pyykkö, Relativistic Effects in Structural Chemistry, *Chem. Rev.*,1988, **88**, 563. A very detailed paper from the main originator of the studies of relativistic effects in chemistry. There are 428 references!

N. Kaltsoyannis, Relativistic Effects in Inorganic and Organometallic Chemistry, *J. Chem. Soc., Dalton Trans.*, 1997, 1. This review paper covers relativistic effects very well, and contains 88 references to the original literature on the basis of the effects and on specific cases.

Further Reading

U. Müller, *Inorganic Structural Chemistry,* 2nd edn., Wiley, New York, 1992

S. E. Dann, *Reactions and Characterization of Solids*, Royal Society of Chemistry, Cambridge, 2000. A companion title in the Tutorial Chemistry Texts series.

Appendix

1   Selected Character Tables

The character tables for the point groups which are relevant to most of the problems and discussions in this book are listed here. The transformation properties of the s, p and d orbitals are indicated by their usual algebraic descriptions appearing in the appropriate row of each table.

C_{2v}	E	C_2	$\sigma_v(xz)$	$\sigma_v(yz)$	
A_1	1	1	1	1	z, x^2, y^2, z^2
A_2	1	1	-1	-1	xy
B_1	1	-1	1	-1	x, xz
B_2	1	-1	-1	1	y, yz

C_{3v}	E	$2C_3$	$3\sigma_v$	
A_1	1	1	1	$z, x^2 + y^2, z^2$
A_2	1	1	-1	
E	2	-1	0	$(x, y)(x^2 - y^2, xy)(xz, yz)$

D_{3h}	E	$2C_3$	$3C_2$	σ_h	$2S_3$	$3\sigma_v$	
A_1'	1	1	1	1	1	1	$x^2 + y^2, z^2$
A_2'	1	1	-1	1	1	-1	
E'	2	-1	0	2	-1	0	$(x, y)(x^2 - y^2, xy)$
A_1''	1	1	1	-1	-1	1	
A_2''	1	1	-1	-1	-1	-1	z
E''	2	-1	0	-2	1	0	(xz, yz)

D_{4h}	E	$2C_4$	C_2	$2C_2'$	$2C_2''$	i	$2S_4$	σ_h	$2\sigma_v$	$2\sigma_d$	
A_{1g}	1	1	1	1	1	1	1	1	1	1	$x^2 + y^2, z^2$
A_{2g}	1	1	1	-1	-1	1	1	1	-1	-1	
B_{1g}	1	-1	1	1	-1	1	-1	1	1	-1	$x^2 - y^2$
B_{2g}	1	-1	1	-1	1	1	-1	1	-1	1	xy
E_g	2	0	-2	0	0	2	0	-2	0	0	(xz, yz)
A_{1u}	1	1	1	1	1	-1	-1	-1	-1	-1	
A_{2u}	1	1	1	-1	-1	-1	-1	-1	1	1	z
B_{1u}	1	-1	1	1	-1	-1	1	-1	-1	1	
B_{2u}	1	-1	1	-1	1	-1	1	-1	1	-1	
E_u	2	0	-2	0	0	-2	0	2	0	0	(x, y)

T_d	E	$8C_3$	$3C_2$	$6S_4$	$6\sigma_d$	
A_1	1	1	1	1	1	$x^2 + y^2 + z^2$
A_2	1	1	1	-1	-1	
E	2	-1	2	0	0	$(2z^2 - x^2 - y^2 , x^2 - y^2)$
T_1	3	0	-1	1	-1	
T_2	3	0	-1	-1	1	$(x, y, z)(xy, xz, yz)$

$C_{\infty v}$	E	C_2	$2C_\infty^\phi$	$\dots$	$\infty\sigma_v$		
Σ^+	1	1	1	$\dots$	1	z	$x^2 + y^2, z^2$
Σ^-	1	1	1	$\dots$	-1		
Π	2	2	$2\cos\phi$	$\dots$	0	(x, y)	(xz, yz)
Δ	2	2	$2\cos2\phi$	$\dots$	0		$(x^2 - y^2, xy)$
Φ	2	2	$2\cos3\phi$	$\dots$	0		
$\dots$	$\dots$	$\dots$	$\dots$	$\dots$	$\dots$		

$D_{\infty h}$	E	$2C_\infty^\phi$	$\dots$	$\infty\sigma_v$	i	$2S_\infty^\phi$	$\dots$	∞C_2	
Σ_g^+	1	1	$\dots$	1	1	1	$\dots$	1	$x^2 + y^2, z^2$
Σ_g^-	1	1	$\dots$	-1	1	1	$\dots$	-1	
Π_g	2	$2\cos\phi$	$\dots$	0	2	$-2\cos\phi$	$\dots$	0	(xz, yz)
Δ_g	2	$2\cos2\phi$	$\dots$	0	2	$2\cos2\phi$	$\dots$	0	$(x^2 - y^2, xy)$
$\dots$	$\dots$	$\dots$	$\dots$	$\dots$	$\dots$	$\dots$	$\dots$	$\dots$	
Σ_u^+	1	1	$\dots$	1	-1	-1	$\dots$	-1	z
Σ_u^-	1	1	$\dots$	-1	-1	-1	$\dots$	1	
Π_u	1	$2\cos\phi$	$\dots$	0	-2	$2\cos\phi$	$\dots$	0	(x, y)
Δ_u	1	$2\cos2\phi$	$\dots$	0	-2	$-2\cos2\phi$	$\dots$	0	
$\dots$	$\dots$	$\dots$	$\dots$	$\dots$	$\dots$	$\dots$	$\dots$	$\dots \dots$	

O_h	E	$8C_3$	$6C_2$	$6C_4$	$3C_2$ $(= C_4^2)$	i	$6S_4$	$8S_6$	$3\sigma_h$	$6\sigma_d$	
A_{1g}	1	1	1	1	1	1	1	1	1	1	$x^2 + y^2 + z^2$
A_{2g}	1	1	−1	−1	1	1	−1	1	1	−1	
E_g	2	−1	0	0	2	2	0	−1	2	0	$(2z^2 - x^2 - y^2,$ $x^2 - y^2)$
T_{1g}	3	0	−1	1	−1	3	1	0	−1	−1	
T_{2g}	3	0	1	−1	−1	3	−1	0	−1	1	(xy, xz, yz)
A_{1u}	1	1	1	1	1	−1	−1	−1	−1	−1	
A_{2u}	1	1	−1	−1	1	−1	1	−1	−1	1	
E_u	2	−1	0	0	2	−2	0	1	−2	0	
T_{1u}	3	0	−1	1	−1	−3	−1	0	1	1	(x, y, z)
T_{2u}	3	0	1	−1	−1	−3	1	0	1	−1	

2 The Reduction of a Representation to a Sum of Irreducible Representations

In the text, when the character of a set of orbitals is deduced to give a reducible representation, the reduction to a sum of irreducible representations has been carried out by inspection of the appropriate character table. In some instances this procedure can be lengthy and unreliable. The formal method can also be lengthy, but it is highly reliable, although not to be recommended for simple cases where inspection of the character table is usually sufficient. The formal method will be explained by doing an example.

The three 1s atomic orbitals of the trigonally pyramidal C_{3v} NH_3 molecule have the character given below:

C_{3v}	E	$2C_3$	$3\sigma_v$
$3 \times N(1s)$	3	0	1

The coefficients of the symmetry elements along the top of the above classification (the same as those across the top of the C_{3v} character table), i.e. 1, 2 and 3, give a total of six which is the *order* of the point group, denoted by h. The relationship used to test the hypothesis that the reducible representation contains a particular irreducible representation is:

$$a(\text{irreducible}) = 1/h \ \Sigma[g \times \chi(R) \times \chi(I)]$$

where g is the number of elements in the class (i.e. the coefficients of the symmetry operations across the top line of the character table), $\chi(R)$ is the character of reducible representation and $\chi(I)$ is the character of the irreducible representation.

In the present case the test for the presence of an A_1 representation depends upon the sum:

$$a(A_1) = 1/6[1 \times 3 \times 1 + 2 \times 0 \times 1 + 3 \times 1 \times 1] = 6/6 = 1$$

implying that the reducible representation contains one A_1 representation.

The test for an A_2 representation is:

$$a(A_2) = 1/6[1 \times 3 \times 1 + 2 \times 0 \times 1 + 3 \times 1 \times -1] = 0/6 = 0$$

so confirming that the irreducible representation does not contain an A_2 representation.

This leaves the possibility that it would contain an E representation, for which the equation is:

$$a(E) = 1/6[1 \times 3 \times 2 + 2 \times 0 \times -1 + 3 \times 1 \times 0] = 6/6 = 1$$

confirming that the reducible representation does contain an E representation. There being no further sums to do in this case, it can now be concluded that the representation of the three 1s orbitals of the hydrogen atoms of the ammonia molecule transform as the sum $A_1 + E$.

Answers to Problems

Chapter 1

1.1. (a) Invalid; l must have a lower value than n. (b) Invalid; if $l = 0$ then m_l can only be zero or if $m_l = -1$ then l should be at least 1. (c) Valid. (d) Valid. (e) Invalid; m_l cannot have a value greater than that of l. (f) Valid.

1.2. The degeneracies of the 3p, 4d and 5d orbitals of the hydrogen atom are 3, 5 and 5, respectively. The degeneracy of a set of orbitals with a specific value of l is given by the number of permitted values of m_l, i.e. $2l + 1$. If $n > 2$, l can have the value 2 and therefore m_l has the five values 2, 1, 0, –1 and –2. This and the Pauli principle ensure that there can be only five d orbitals for a given n value, providing that $n > 2$.

Chapter 2

2.1. The point groups to which the examples belong are: C_s (n); $C_{\infty v}$ (a) and (h); C_{2v} (d), (e), (f), (g), (i), (j), (k) and (m); C_{3v} (l) and (p); $D_{\infty h}$ (b) and (c); T_d (r).

Chapter 4

4.4. The charge separation in HCl is given by:

$$Q = \frac{1.03 \times 3.336 \times 10^{-30}}{128 \times 10^{-12} \times 1.6 \times 10^{-19}} = 0.168e$$

i.e. the molecule is 16.8% ionic. The differences in electronegativity coefficients in HF and HCl are, respectively, 2.0 and 0.9 and HCl would be expected to be less ionic than HF.

4.5. The anomalous dipole moment of CO is due to the occupation of the non-bonding $5\sigma^+$ orbital, which is localized on the carbon atom (and is used to donate charge to a metal centre when CO acts as a ligand).

Chapter 5

5.2. The ground state of the beryllium atom is zero-valent $2s^2$. This may be converted to the divalent state $2s^1 2p^1$ by making use of the $2p_z$ orbital. The addition of the two electrons from the ligand fluorine atoms makes two electron pairs which make the molecule linear. The BeF_2 molecule belongs to the $D_{\infty h}$ point group so the appropriate diagram to use for its electronic configuration is Figure 5.14. The electronic configuration of the molecule is $(4\sigma_g^+)^2(3\sigma_u^+)^2(1\pi_u)^4(1\pi_g)^4$, the first two orbitals are σ bonding, the $1\pi_u$ are also bonding, and the $1\pi_g$ are non-bonding and localized on the fluorine ligand atoms. The bond order of the two Be–F bonds is two. The valence bond equivalent of the molecular bonding is shown in Figure A1. There are two coordinate bonds with electron pairs donated by the fluorine atoms to complete the octet in the valence shell of the beryllium atom.

$$F \rightleftharpoons Be \rightleftharpoons F$$

Figure A1

CS_2 (same valence configuration as CO_2) makes use of the four-valent state of the carbon atom $2s^1 2p^3$ and with the two electrons from both the ligand sulfur atoms has four pairs of valence electrons. Two of these are σ-type and make the molecule linear. The diagram of Figure 5.14 is used to write the electronic configuration as $(4\sigma_g^+)^2(3\sigma_u^+)^2(1\pi_u)^4(1\pi_g)^4$, the first three orbitals being bonding and making the bond order of each C–S bond equal to 2.

The ClF_2^+ ion has a central chlorine atom which may be considered to be singly ionized Cl^+ which would be divalent, similar to the ground state of S. This $3s^2 3p^4$ configuration then accepts two electrons from the ligand fluorine atoms to produce four electron pairs, two of which are bonding and the other two non-bonding. The shape is dependent upon a tetrahedral distribution of the four electron pairs and would lead to a bent ion with a bond angle somewhat less than the regular tetrahedral angle. MO treatment would use Figure 5.20 and the accommodation of the 20 valence electrons (six from Cl^+ and seven each from the fluorine atoms) would be best in the bent shape using orbitals on the right-hand side of the diagram. The strongly angle-dependent $6a_1$ orbital would be occupied and that would stabilize the bent ion.

5.3. The chlorine dioxide molecule, ClO_2, contains a chlorine atom in the unusual formal oxidation state (+4), and is an *odd* molecule containing 19 valency electrons. The molecule has a bond angle of 117° with a Cl–O bond length of 148 pm. Apply the VSEPR treatment to the molecule, and use Figure 5.20 to decide the MO electronic configuration. Compare the results from the two theories in their ability to explain the bond angle and the bond length. The single bond covalent radii of Cl and O are 99 and 74 pm, respectively.

The chlorine dioxide molecule can be treated by the VSEPR theory by converting the uni-valent ground state of the central chlorine atom, $3s^23p^5$, into a five-valent state by promoting two of the 3p electrons to 3d orbitals: $3s^23p^33d^2$. The acquisition of two electrons from each of the ligand divalent oxygen atoms gives the configuration $3s^23p^63d^3$, which corresponds to five pairs of electrons, two of which are σ and two are π orbitals. That leaves a non-bonding pair and a single non-bonding electron. The shape would be based on a roughly tetrahedral disposition of the 3½ σ pairs. The single electron in one of the tetrahedral positions would allow the O–C–O bond angle to be somewhat larger than the regular tetrahedral angle. The shorter bond length, compared to a single bond between chlorine and oxygen, implies that the bond order is greater than one, as would be expected for the divalent oxygen.

MO treatment would use Figure 5.20 and the accommodation of the 19 valence electrons (seven from Cl and six each from the oxygen atoms) would be best in the bent shape using orbitals on the right-hand side of the diagram. The strongly angle-dependent $6a_1$ orbital would be occupied and that would stabilize the bent ion.

Chapter 6

6.2. To apply VSEPR theory to the carbonate ion it is essential to convert the central carbon atom into its four-valent state $2s^12p^3$ and to feed into that configuration two electrons from a ligand oxygen atom to make a σ + π bonding arrangement and two electrons from the two O^- ligands. That completes the four pairs of valency electrons for the carbon atom. Only three of the pairs are σ-type and so the predicted shape is a trigonal plane. The given data imply that the C–O bonds have an order greater than one and this is understandable in terms of the three canonical forms shown in Figure A2 and from the initial assumption that there is a π bond present. The bond order, therefore, is 1⅓ for each of the three bonds, the π bond being shared between them.

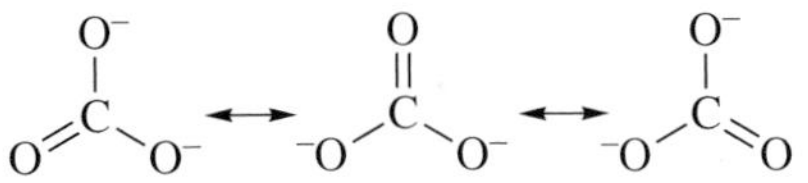

Figure A2

With the VSEPR prediction in mind, the most appropriate MO diagram to use is that for BF_3 in Figure 6.6. The BF_3 molecule is isoelectronic with the carbonate ion and similarities in structure and bonding are to be expected. The three σ bonding orbitals $3a_1'$ and $2e'$ are occupied, as is the π-type $1a_2''$ orbital. The latter is delocalized over the carbonate ion and reduces the C–O distances compared to that expected for a single bond.

6.3. The ground state of the sulfur atom is $3s^2 3p^4$ and all six electrons must be unpaired to produce the six-valent state necessary for the formation of SO_3, each oxygen atom making a supposed double bond with the central S atom. This means promoting one of the 3s electrons and one of the 3p electrons to separate 3d orbitals to give the configuration $3s^1 3p^3 3d^2$, which can then accept the six electrons from the three ligand oxygen atoms to give a valence shell with six electron pairs, three of which are σ pairs and three which are π pairs. The molecule will be a trigonal plane. The evidence for the double bonding comes from the S–O distances, which are considerably shorter than those expected for single bonds (178 pm).

   The MO treatment of SO_3 would make use of a diagram such as Figure 6.6 modified by incorporating the two sulfur 3d orbitals that could take part in the trigonally planar bonding as π orbitals, i.e. the $3d_{xz}$ and $3d_{yz}$ orbitals which have e'' symmetry. The extra e'' bonding orbitals would accommodate four bonding π electrons, which would stabilize the molecule and ensure that the S–O bond order was 2.

6.4. The H_2F^+ ion is isoelectronic with the water molecule and would be expected to have the same shape: bent with an angle somewhat smaller than the regular tetrahedral angle. The fluorine atom could be ionized to give a positively charged F^+ ion which would then have a two-valent $2s^2 2p^4$ configuration. This could accept the two electrons from the ligand hydrogen atoms to give four pairs of valence electrons. These would be disposed tetrahedrally and the two bonding pairs squeezed together somewhat by the greater repulsion from the two non-bonding pairs. Figure 5.12 may be used to verify that the electronic configurations of H_2F^+ and H_2O should be identical.

The SbF_6^- ion has a central antimony atom with a ground state of $5s^25p^3$ which is trivalent. If the 5s electrons were to be unpaired and one promoted to a 5d orbital the antimony would then be five-valent. If it accepted an electron to make it Sb^- and the electron was accommodated in another 5d orbital the ion would be six-valent and could accept the six electrons from the six ligand fluorine atoms to give a valence shell of six σ pairs. This would indicate that the ion should have a regular octahedral shape.

A σ-only MO treatment of SbF_6^- is carried out according to the general protocol described in Chapter 4. The orbitals of the central antimony atom transform according to the O_h character table as 5s (a_{1g}), 5p (t_{1u}), $5d_{xy}$, $5d_{xz}$ and $5d_{yz}$ (t_{2g}), $5d_{z^2}$ and $5d_{x^2-y^2}$ (e_g). The six σ-type ligand orbitals may be arranged so that they are in a regular octahedral arrangement directed at the central antimony atom, as shown in Figure A3.

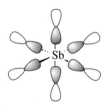

Figure A3

They form a reducible representation with the characters:

	E	C_3	C_2	C_4	C_2	i	S_4	S_6	σ_h	σ_d
$6 \times \sigma(F)$	6	0	0	2	2	0	0	0	4	2

This conclusion is based on the number of fluorine σ orbitals unaffected by the various symmetry operations. The C_3 axes pass through the triangular faces of the regular tetrahedron. The first C_2 axes bisect any two of the Cartesian axes and rotation around any of these affects the positions of all six ligand orbitals. The C_4 axes coincide with the Cartesian axes and rotation around any one of them affects four orbitals, leaving two unaffected. The C_4 axes are also C_2 axes and the rotation around any one of them leaves two orbitals unaffected. The inversion operation affects all six orbitals. The S_4 and S_6 operations affect all six orbitals. Reflexion in a horizontal plane leaves four orbitals unaffected and reflexion in any of the dihedral planes (those passing through the central atom and containing the C_2 axes that bisect any two of the Cartesian axes) leaves two orbitals unaffected. The representation is resolved into irreducible representations as shown by the sum:

	E	C_3	C_2	C_4	C_2	i	S_4	S_6	σ_h	σ_d
a_{1g}	1	1	1	1	1	1	1	1	1	1
e_g	2	-1	0	0	2	2	0	-1	2	0
t_{1u}	3	0	-1	1	-1	-3	-1	0	1	1
$6 \times \sigma(F)$	6	0	0	2	2	0	0	0	4	2

so that $6 \times \sigma(F) = a_{1g} + e_g + t_{1u}$. A MO diagram is shown in Figure A4 in which the sulfur orbitals and those of the fluorine σ group orbitals of the same symmetries are allowed to interact to give bonding and anti-bonding MOs.

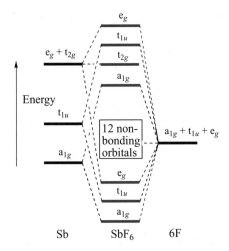

Figure A4

The 12 σ electrons (*i.e.* one negative charge plus five from Sb and one from each fluorine atom) would occupy the six bonding orbitals to give each Sb–F bond an order of one. 5d orbital participation is built into the simplified bonding scheme and it possible that the t_{2g} antimony orbitals would participate in π-type bonding with suitable ligand orbitals. No anti-bonding orbitals are occupied so it would seem that a regular octahedron is the appropriate symmetry for SbF_6^-.

The TeF_6^{2-} ion according to VSEPR theory should be a distorted octahedron since the valence shell has seven of pairs of σ electrons. Experimentally the ion is a regular octahedron, indicating that the extra electrons are not 'stereochemically active'. In MO theory the extra pair is accommodated by the anti-bonding a_{1g} orbital and since this is completely symmetric it is not expected to affect the shape of the ion. The isoelectronic XeF_6 molecule is 'fluxional' between structures that seem to have the 'lone pair' either capping one of the triangular faces of the octahedron or occupying a position in what would be a pentagonal plane. This could be due to the accommodation of the highest energy pair of electrons in an anti-bonding level other than the completely symmetric a_{1g} orbital which is used in TeF_6^{2-}.

6.6. The silicon atom is formally divalent and may be converted to its four-valent state by the promotion of a 3s electron to the vacant 3p orbital: $3s^1 3p^3$. This can then accept the four electrons from the ligand fluorine atoms to give four pairs of σ-bonding electron pairs. The molecule is predicted to be tetrahedral. The sulfur atom is formally divalent and can be converted to its four-valent state by promoting one of the paired 3p electrons to a 3d orbital. This state can then accept the four electrons from the ligand fluorine atoms to give four pairs of σ-bonding electron pairs plus one non-bonding pair. The shape will be dependent upon a trigonally bipyramidal disposition of the five pairs of electrons. The non-bonding pair should occupy the trigonal plane, thus giving the distorted tetrahedral shape of C_{2v} symmetry known as the see-saw or butterfly shape.

The electronic configuration of a regularly tetrahedral SF_4 molecule would be $(2a_1)^2 (1t_2)^6 (3a_1)^2$ [making use of Figure 6.3 and retaining the Mulliken numbering for carbon] with the $3a_1$ orbital containing a pair of anti-bonding electrons. If a distortion to C_{2v} symmetry occurs, the anti-bonding electrons would occupy an a_1 orbital which is stabilized by interaction with the a_1 orbital produced from the t_2 orbitals in the lower symmetry. They form a_1, b_1 and b_2 orbitals and the a_1 orbital is destabilized. The occupied a_1 orbital is the non-bonding sp hybrid in the third position of the trigonal plane, the other two positions being filled by two ligand fluorine atoms. The general principle holds in this example.

Chapter 7

7.1. The lattice energy of calcium oxide (Born–Landé) is 3554 kJ mol^{-1}, the enthalpy of lattice formation is $-3554 - 6 = -3560$ kJ mol^{-1}. The calculated standard enthalpy of formation of calcium oxide is -677 kJ mol^{-1}, about 7% more exothermic than the observed value, probably due to the inherent error in the calculation.

7.2. The lattice energy of zinc sulfide (Born–Landé) is 3136 kJ mol^{-1}, the enthalpy of lattice formation is $-3136 - 6 = -3142$ kJ mol^{-1}. The calculated standard enthalpy of formation of zinc sulfide is 187 kJ mol^{-1}, clearly at odds with the ionic model. The solid is considerably covalent.

7.3. The enthalpy change in the formation of N_4 from $2N_2$ is 910 kJ mol^{-1}, that for the formation of P_4 from $2P_2$ is -70 kJ mol^{-1}. Thus, the stable forms of the elements are N_2 and P_4.

Subject Index